Handbook of
APPLIED CIVIL ENGINEERING DESIGN

Handbook of APPLIED CIVIL ENGINEERING DESIGN

Alice Walker
Shwan Root

KNOWLEDGE BAKERS

Cataloging in Publication Data--DK
Courtesy: D.K. Agencies (P) Ltd. <docinfo@dkagencies.com>

Walker, Alice *(Writer on civil engineering)*, **author.**
Handbook of applied civil engineering design / Alice Walker, Shwan Root.
pages cm
Includes bibliographical references and index.
ISBN 9789391502348

1. Civil engineering. 2. Engineering design. I. Root, Shwan, author. II. Title.

LCC TC145.W35 2023 | DDC 627 23

ISBN: 978-93-91502-34-8

9 789391 502348

Published by : **Knowledge Bakers**
Pune, India
Email: knowledgebakers2019@gmail.com

Digitally Printed at : **Replika Press Pvt. Ltd.**

Preface

Traditionally, consultants design and contractors construct. Both Parties are employed by the client through separate contracts. Thus the designer has a direct relationship with the client and is focused on the client's needs. D&C requires a different approach and both contractors and consultants can find it difficult to adjust to this method of procurement. Consultants may be reluctant to work with a contractor to produce a suitable design and contractors often struggle with the implications of being responsible for a design they did not produce. Similarly, clients are often left feeling exposed, as they no longer have direct control over the design.

Many of the problems in D&C arise from a lack of suitably qualified and experienced managers on both sides of the designer-constructor interface. These problems can be made worse if the client insists on an overly bureaucratic regime of checks and approvals. The importance of this interface is emphasized throughout the guide. There must be an individual with responsibility for managing this interface. Involving the constructor, the operator and the eventual end-user in the initial stages of design work can bring the client significant benefits, from single point accountability through to improved operational and maintenance characteristics. Project teams can benefit from more direct communication, resulting in greater awareness of different aspects of a project and fewer errors due to misinterpretation and misunderstanding. There must, however, be genuine teamwork based on respect for each other's interests, expertise and responsibility.

With the civil engineering industry, D&C is increasingly being adopted as the preferred procurement route. It is, however, far from new. The UK building industry has a long established record of successfully combining the skills of designers and constructors. Many manufacturing companies are founded on the principle of designers and constructors teaming up to create a product that meets customers' requirements and expectations. A cultural change is taking place in many industries and there are lessons to be learned from those trying to introduce change. Examples are the UK offshore oil and gas industry's CRINE initiative, an abbreviation for "Cost Reduction Initiative for the New Era" or the chemical and process industry's ACTIVE, "Achieving Competitiveness Through Innovation and Value Engineering".

Those involved in CRINE – and it has very wide support in the oil and gas industry–freely admit that many of the ideas in the initiative are not new. The purpose of CRINE is to provide "the tools to enable the UK Offshore oil and gas industry to reduce costs by at least 30 per cent" and this is necessary for the industry to survive. CRINE focuses on enhancing trust and cooperation throughout the industry, enabling the objectives of owners and contractors to be realized to create a "win-win" scenario, where costs can be cut and profits can be increased. CRINE is achieving results, according to the industry, because its need is recognized. The purpose of ACTIVE is very similar to CRINE and because its need is recognized. The purpose of ACTIVE is very similar to CRINE and both have been signed up to by some of the UK's largest civil engineering contractors.

The present book has been designed to outline the basic and fundamental aspects of this subject to be understood in its right perspective. The book uses straight forward, less-technical jargon and manages to introduce each chapter with a basic concept, which ultimately evolves into a more specific detailed principle.

Dr Alice Walker
Dr Shwan Root

Contents

Chapter 1

Introduction to Civil Engineering

Civil engineering is a professional engineering discipline that deals with the design, construction, and maintenance of the physical and naturally built environment, including works like roads, bridges, canals, dams, and buildings.

Figure: *Design of complex structures like the International Space Station needs in-depth understanding of Structural Analysis.*

Civil engineering is the oldest engineering discipline after military engineering, and it was defined to distinguish non-military engineering from military engineering. It is traditionally broken into several sub-disciplines including environmental engineering, geotechnical engineering, structural engineering, transportation engineering, municipal or urban engineering, water resources engineering, materials engineering, coastal engineering, surveying, and construction engineering. Civil engineering takes place on all levels: in the public sector from municipal through to national governments, and in the private sector from individual homeowners through to international companies.

History of Structural Engineering

The history of structural engineering dates back to at least 2700 BC when the step pyramid for Pharaoh Djoser was built by Imhotep, the first engineer in history known by name. Pyramids were the most common major structures built by ancient civilizations because it is a structural form which is inherently stable and can be almost infinitely scaled (as opposed to most other structural forms, which cannot be linearly increased in size in proportion to increased loads).

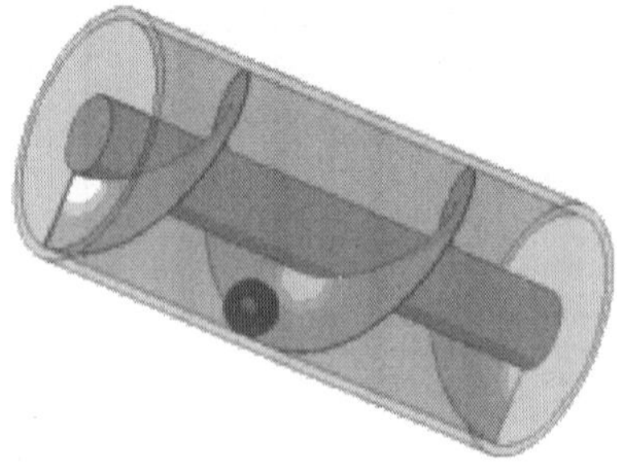

Figure: *The Archimedes screw was operated by hand and could raise water efficiently.*

Another notable engineering feat from antiquity still in use today is the qanat water management system. Qanat technology developed in the time of the Medes, the predecessors of the Persian Empire (modern-day Iran which has the oldest and longest Qanat (older than 3000 years and longer than 71 km) that also spread to other cultures having had contact with the Persian.

Throughout ancient and medieval history most architectural design and construction was carried out by artisans, such as stone masons and carpenters, rising to the role of master builder. No theory of structures existed and understanding of how structures stood up was extremely limited, and based almost entirely on empirical evidence of 'what had worked before'. Knowledge was retained by guilds and seldom supplanted by advances. Structures were repetitive, and increases in scale were incremental.

Figure: *Pont du Gard, France, a Roman aqueduct built circa 19 BC.*

No record exists of the first calculations of the strength of structural members or the behaviour of structural material, but the profession of structural engineer only really took shape with the industrial revolution and the re-invention of concrete. The physical sciences underlying structural engineering began to be understood in the Renaissance and have been developing ever since.

Early Structural Engineering Developments

The recorded history of structural engineering starts with the ancient Egyptians. In the 27th century BC, Imhotep was the first structural engineer known by name and constructed the first known step pyramid in Egypt. In the 26th century BC, the Great Pyramid of Giza was constructed in Egypt. It remained the largest man-made structure for millennia and was considered an unsurpassed feat in architecture until the 19th century AD.

The understanding of the physical laws that underpin structural engineering in the Western world dates back to the 3rd century BC, when Archimedes published his work *On the Equilibrium of Planes* in two volumes, in which he sets out the *Law of the Lever*, stating:

> *"Equal weights at equal distances are in equilibrium, and equal weights at unequal distances are not in equilibrium but incline towards the weight which is at the greater distance."*

Archimedes used the principles derived to calculate the areas and centres of gravity of various geometric figures including triangles, paraboloids, and hemispheres. Archimedes work on this and his work on calculus and geometry, together with Euclidean geometry, underpin much of the mathematics and understanding of structures in modern structural engineering. The ancient Romans made great bounds in structural engineering, pioneering large structures in masonry and concrete, many of which are still standing today. They include aqueducts, thermae, columns, lighthouses, defensive walls and harbours. Their methods are recorded by Vitruvius in his De Architecture written in 25 BC, a manual of civil and structural engineering with extensive sections on materials and machines used in construction. One reason for their success is their accurate surveying techniques based on the dioptre, groma and chorobates.

Centuries later, in the 15th and 16th centuries and despite lacking beam theory and calculus, Leonardo da Vinci produced many engineering designs based on scientific observations and rigour, including a design for a bridge to span the Bosporus. Though dismissed at the time, the design has since been judged to be both feasible and structurally valid.

The foundations of modern structural engineering were laid in the 17th century by Galileo Galilei, Robert Hooke and Isaac Newton with the publication of three great scientific works. In 1638 Galileo published *Dialogues Relating to Two New Sciences*, outlining the sciences of the strength of materials and the motion of objects (essentially defining gravity as a force giving rise to a constant acceleration). It was the first establishment of a scientific approach to structural engineering, including the first attempts to develop a theory for beams. This is also regarded as the beginning of structural analysis, the mathematical representation and design of building structures. This was followed in 1676 by Robert Hooke's first statement of Hooke's Law, providing a scientific understanding of elasticity of materials and their behaviour under load.

Eleven years later, in 1687, Sir Isaac Newton published *Philosophiae Naturales Principle Mathematics*, setting out his Laws of Motion, providing for the first time an understanding of the fundamental laws governing structures. Also in the 17th century, Sir Isaac Newton and Gottfried Leibniz both independently developed the Fundamental theorem of calculus, providing one of the most important mathematical tools in engineering.

Further advances in the mathematics needed to allow structural engineers to apply the understanding of structures gained through the work of Galileo, Hooke and Newton during the 17th century came in the 18th century when Leonhard Euler pioneered much of the mathematics and many of the methods which allow structural engineers to model and analyse structures. Specifically, he developed the Euler-Bernoulli beam equation with Daniel Bernoulli (1700–1782) circa 1750 - the fundamental theory underlying most structural engineering design. Daniel Bernoulli, with Johann (Jean) Bernoulli (1667–1748), is also credited with formulating the theory of virtual work, providing a tool using equilibrium of forces and compatibility of geometry to solve structural problems. In 1717 Jean Bernoulli wrote to Pierre Varignon explaining the principle of virtual work, while in 1726 Daniel Bernoulli wrote of the "composition of forces".

In 1757 Leonhard Euler went on to derive the Euler buckling formula, greatly advancing the ability of engineers to design compression elements.

Modern Developments in Structural Engineering

Throughout the late 19th and early 20th centuries, materials science and structural analysis underwent development at a tremendous pace.

Though elasticity was understood in theory well before the 19th century, it was not until 1821 that Claude-Louis Navier formulated the general theory of elasticity in a mathematically usable form. In his *leçons* of 1826 he explored a great range of different structural theory, and was the first to highlight that the role of a structural engineer is not to understand the final, failed state of a structure, but to prevent that failure in the first place. In 1826 he also established the elastic modulus as a property of materials independent of the second moment of area, allowing engineers for the first time to both understand structural behaviour and structural materials.

Towards the end of the 19th century, in 1873, Carlo Alberto Castigliano presented his dissertation "Intorno ai sistemi elastici", which contains his theorem for computing displacement as partial derivative of the strain energy.

In 1824, Portland cement was patented by the engineer Joseph Aspdin as *"a superior cement resembling Portland Stone"*, British Patent no. 5022. Although different forms of cement already existed (Pozzolanic cement was used by the Romans as early as 100 B.C. and even earlier by the ancient Greek and Chinese civilizations) and were in common usage in Europe from the 1750's, the discovery made by Aspdin used commonly available, cheap materials, making concrete construction an economical possibility.

Developments in concrete continued with the construction in 1848 of a rowing boat built of ferrocement - the forerunner of modern reinforced concrete - by Joseph-Louis Lambot. He patented his system of mesh reinforcement and concrete in 1855, one year after W.B. Wilkinson also patented a similar system. This was followed in 1867 when a reinforced concrete planting tub was patented by Joseph Monier in Paris, using steel mesh reinforcement similar to that used by Lambot and Wilkinson. Monier took the idea forward, filing several patents for tubs, slabs and beams, leading eventually to the Monier system of reinforced structures, the first use of steel reinforcement bars located in areas of tension in the structure.

Steel construction was first made possible in the 1850's when Henry Bessemer developed the Bessemer process to produce steel. He gained patents for the process in 1855 and 1856 and successfully completed the conversion of cast iron into cast steel in 1858. Eventually mild steel would replace both wrought iron and cast iron as the preferred metal for construction.

During the late 19th century, great advancements were made in the use of cast iron, gradually replacing wrought iron as a material of choice. Ditherington Flax Mill in Shrewsbury, designed by Charles Bage, was the first building in the world with an interior iron frame. It was built in 1797. In 1792 William Strutt had attempted to build a fireproof mill at Belper in Derby (Belper West Mill), using cast iron columns and timber beams within the depths of brick arches that formed the floors. The exposed beam soffits were protected against fire by plaster. This mill at Belper was the world's first attempt to construct fireproof buildings, and is the first example of fire engineering. This was later improved upon with the construction of Belper North Mill, a collaboration between Strutt and Bage, which by using a full cast iron frame represented the world's first "fire proofed" building.

The Forth Bridge was built by Benjamin Baker, Sir John Fowler and William Arrol in 1889, using steel, after the original design for the bridge by Thomas Bouch was rejected following the collapse of his Tay Rail Bridge. The Forth Bridge was one of the first major uses of steel, and a landmark in bridge design. Also in 1889, the wrought-iron Eiffel Tower was built by Gustave Eiffel and Maurice Koechlin, demonstrating the potential of construction using iron, despite the fact that steel construction was already being used elsewhere.

During the late 19th century, Russian structural engineer Vladimir Shukhov developed analysis methods for tensile structures, thin-shell structures, lattice shell structures and new structural geometries such as hyperboloid structures. Pipeline transport was pioneered by Vladimir Shukhov and the Branobel company in the late 19th century.

Again taking reinforced concrete design forwards, from 1892 onwards François Hennebique's firm used his patented reinforced concrete system to build thousands of structures throughout Europe. Thaddeus Hyatt in the US and Wayss & Freitag in Germany also patented systems. The firm *AG für Monierbauten* constructed 200 reinforced concrete bridges in Germany between 1890 and 1897 The great pioneering uses of reinforced concrete however came during the first third of the 20th century, with Robert Maillart and others furthering of the understanding of its behaviour. Maillart noticed that many concrete bridge structures were significantly cracked, and as a result left the cracked areas out of his next bridge design - correctly believing that if the concrete was cracked, it was not contributing to the strength. This resulted in the revolutionary Salginatobel Bridge design. Wilhelm Ritter formulated the truss theory for the shear design of reinforced concrete beams in 1899, and Emil Mörsch improved this in 1902. He went on to demonstrate that treating concrete in compression as a linear-elastic material was a conservative approximation of its behaviour. Concrete design and analysis has been progressing ever since, with the development of analysis methods such as yield line theory, based on plastic analysis of concrete (as opposed to linear-elastic), and many different variations on the model for stress distributions in concrete in compression. Prestressed concrete, pioneered by Eugène Freyssinet with a patent in 1928, gave a novel approach in overcoming the weakness of concrete structures in tension. Freyssinet constructed an experimental prestressed arch in 1908 and later used the technology in a limited form in the Plougastel Bridge in France in 1930. He went on to build six prestressed concrete bridges across the Marne River, firmly establishing the technology.

Structural engineering theory was again advanced in 1930 when Professor Hardy Cross developed his Moment distribution method, allowing the real stresses of many complex structures to be approximated quickly and accurately.

In the mid 20th century John Fleetwood Baker went on to develop the plasticity theory of structures, providing a powerful tool for the safe design of steel structures.

High-rise construction, though possible from the late 19th century onwards, was greatly advanced during the second half of the 20th century. Fazlur Khan designed structural systems that remain fundamental to many modern high rise constructions and which he employed in his structural designs for the John Hancock Centre in 1969 and Sears Tower in 1973. Khan's central innovation in skyscraper design and construction was the idea of the "tube" and "bundled tube" structural systems for tall buildings. He defined the framed tube structure as "a three dimensional space structure composed of three, four, or possibly more frames, braced frames, or shear walls, joined at or near their edges to form a vertical tubelike structural system capable of resisting lateral forces in any direction by cantilevering from the foundation." Closely spaced interconnected exterior columns form the tube. Horizontal loads, for example wind, are supported by the structure as a whole. About half the exterior surface is available for windows. Framed tubes allow fewer interior columns, and so create more usable floor space. Where larger openings like garage doors are required, the tube frame must be interrupted, with transfer girders used to maintain structural integrity. The first building to apply the tube-frame construction was in the DeWitt-Chestnut Apartment Building which Khan designed in Chicago. This laid the foundations for the tube structures used in most later skyscraper constructions, including the construction of the World Trade Centre.

Another innovation that Fazlur Khan developed was the concept of X-bracing, which reduced the lateral load on the building by transferring the load into the exterior columns. This allowed for a reduced need for interior columns thus creating more floor space, and can be seen in the John Hancock Centre. The first sky lobby was also designed by Khan for the John Hancock Centre in 1969. Later buildings with sky lobbies include the World Trade Centre, Petronas Twin Towers and Taipei 101.

In 1987 Jörg Schlaich and Kurt Schafer published the culmination of almost ten years of work on the strut and tie method for concrete analysis - a tool to design structures with discontinuities such as corners and joints, providing another powerful tool for the analysis of complex concrete geometries.

In the late 20th and early 21st centuries the development of powerful computers has allowed finite element analysis to become a significant tool for structural analysis and design. The development of finite element programs has led to the ability to accurately predict the stresses in complex structures, and allowed great advances in structural engineering design and architecture. In the 1960's and 70's computational analysis was used in a significant way for the first time on the design of the Sydney Opera House roof. Many modern structures could not be understood and designed without the use of computational analysis. Developments in the understanding of materials and structural behaviour in the latter part of the 20th century have been significant, with detailed understanding being developed of topics such as fracture mechanics, earthquake engineering, composite materials, temperature effects on materials, dynamics and vibration control, fatigue, creep and others. The depth and breadth of knowledge now available in structural engineering, and the increasing range of different structures and the increasing complexity of those structures has led to increasing specialisation of structural engineers.

A civil engineer is a person who practices civil engineering; the application of planning, designing, constructing, maintaining, and operating infrastructures while protecting the public and environmental health, as well as improving existing infrastructures that have been neglected. Originally, a civil engineer worked on public works projects and was contrasted with the military engineer, who worked on armaments and defences. Over time, various branches of engineering have become recognised as distinct from civil engineering, including chemical engineering, mechanical engineering, and electrical engineering, while much of military engineering has been absorbed by civil engineering. In some places, a civil engineer may perform land surveying; in others, surveying is limited to construction surveying, unless an additional qualification is obtained. On some U.S. Military bases, the personnel responsible for building and grounds maintenance, such as grass mowing, are called civil engineers and are not required to meet any minimum educational requirements.

Specialisation

Civil engineers usually practice in a particular speciality, such as construction engineering, geotechnical engineering, structural engineering, land development, transportation engineering, hydraulic engineering, and environmental engineering. Some civil engineers, particularly those working for government agencies, may practice across multiple specialisation, particularly when involved in critical infrastructure development or maintenance.

Education and Licensing

In most countries, a civil engineer will have graduated from a post-secondary school with a degree in civil engineering, which requires a strong background in mathematics and the physical sciences; this degree is typically a bachelor's degree, though many civil engineers study further to obtain masters, engineer, doctoral and post doctoral degrees. In many countries, civil engineers are subject to licensure. In jurisdictions with mandatory licensing, people who do not obtain a license may not call themselves "civil engineers."

Europe

Belgium: In Belgium, *Civil Engineer* (abbreviated ir.) (French: *Ingénieur Civil*, Dutch: *Burgerlijk Ingenieur*) is a legally protected title applicable to graduates of the five-year engineering course of one of the six universities and the Royal Military Academy. Their speciality can be all fields of engineering: civil, structural, electrical, mechanical, chemical, physics and even computer science. This use of the title may cause confusion to the English speaker as the Belgian "civil" engineer can have a speciality other than civil engineering. In fact, Belgians use the adjective "civil" as an opposition to military engineers.

The formation of the civil engineer has a strong mathematical and scientific base and is more theoretical in approach than the practical oriented industrial engineering. Educated in a five-year program at a polytechnic. Traditionally, students were required to pass an entrance exam on mathematics to start civil engineering studies. This exam was abolished in 2004 for the Flemish Community, but is still organised in the French Community.

Scandinavia

In Scandinavian countries, civil engineer (civilingenjör (Swedish), sivilingeniør (Norwegian), civilingeniør (Danish)) is a first professional degree, approximately equivalent to Master of Science in Engineering, and a protected title granted to students by selected institutes of technology. As in English the word has its origin in the distinction between civilian and military engineers, as in before the start of the 19th century only military engineers existed and the prefix "civil" was a way to separate those who had studied engineering in a regular University from their military counterparts. Today the degree spans over all fields within engineering, like civil engineering, computer science, electronics engineering, etc.

There is generally a slight difference between a Master of Science in Engineering degree and the Scandinavian civil engineer degree, the latter's programme having closer ties with the industry's demands. A civil engineer is the most well-known of the two; still, the area of expertise remains obfuscated for most of the public. A noteworthy difference is the mandatory courses in mathematics and physics, regardless of the equivalent master degree, e.g. computer science. Although a 'college engineer' (högskoleingenjör, diplomingenjör/ mellaningenjör (Swedish), høgskoleingeniør (Norwegian), diplomingeniør (Danish)) is roughly equivalent to a Bachelor of Science in Scandinavia, to become a 'civil engineer' one often has had to do up to one extra year of overlapping studies compared to attaining a B.Sc./M.Sc. combination. This is because the higher educational system is not fully adopted to the international standard graduation system, since it is treated as a professional degree. Today (2009) this is starting to change due to the Bologna process.

A Scandinavian "civilingenjör" will in international contexts commonly call herself "Master of Science in Engineering" and will occasionally wear an engineering class ring. At the Norwegian Institute of Technology (now the Norwegian University of Science and Technology), the tradition with a NTH Ring goes back to 1914, before the Canadian iron ring. In Norway the title "Sivilingeniør" will no longer be issued after 2007, and have been replaced with "Master i technology". In the English translation of the diploma, the title will be "Master of Science", since "Master of Technology" is not an established title in the English-speaking world. The extra overlapping year of studies have also been abolished with this change to make Norwegian degrees more equal to their international counterparts.

Spain

In Spain, a civil engineering degree can be obtained after four years of study in the various branches of mathematics, physics, mechanics, etc. The earned degree is called *Grado en Ingeniería Civil*. Further studies at a Graduate school include master's and doctoral degrees. Before the current situation, that is, before the implementation of Bologna Process in 2010, a Civil Engineering degree in Spain could be obtained after three or five years of study. In the first case, the earned degree was called *Ingeniero Técnico de Obras Públicas* (*ITOP*), literally translated as "Public Works Engineer"; at the second case, the academic earned degree was called *Ingeniero de Caminos, Canales y Puertos* (often shortened to *Ingeniero de Caminos* or *ICCP*), that literally means "Roads, Canals and Harbours Engineer", though civil engineers in Spain practice in the same fields as civil engineers do elsewhere.

The first Spanish Civil Engineering School was the *Escuela Especial de Ingenieros de Caminos y Canales* (now called *Escuela*

Técnica Superior de Ingenieros de Caminos, Canales y Puertos), established in 1802 in Madrid, followed by the *Escuela Especial de Ayudantes de Obras Públicas* (now called *Escuela Universitaria de Ingeniería Técnica de Obras Públicas de la Universidad Politécnica de Madrid*), founded in 1854 in Madrid. Both schools now belong to the Technical University of Madrid.

In Spain, a Civil Engineer has the technical and legal ability to design projects of any branch, so any Spanish Civil Engineer can oversee projects about structures, buildings (except residential structures which are reserved for architects), foundations, hydraulics, the environment, transportation, urbanism, etc. In Spain, Mechanical and Electrical engineering tasks are included under the Industrial engineering degree.

United Kingdom

A chartered civil engineer (known as certified or professional engineer in other countries) is a member of the Institution of Civil Engineers, and has also passed chartership exams. However a non-chartered civil engineer may be a member of the Institution of Civil Engineers or the Institution of Civil Engineering Surveyors. The description "Civil Engineer" is not restricted to members of any particular professional organisation although "Chartered Civil Engineer" is.

North America

United States: In the United States, civil engineers are typically employed by municipalities, construction firms, consulting engineering firms, architect/engineer firms, state governments, and the federal government. Each State requires engineers who offer their services to the public to be licensed by the State. Licensure is obtained by meeting specified education, examination, and work experience requirements. Specific requirements vary by State. Typically licensed engineers must graduate from an ABET-accredited University or College engineering program, pass the Fundamentals of Engineering exam, obtain several years of engineering experience under the supervision of a licensed engineer, then pass the Principles and Practice of Engineering Exam. After completing these steps and the granting of licensure by a State Board, engineers may use the title "Professional Engineer" or PE in advertising and documents.

Professional Associations

ASCE: The ASCE (American Society of Civil Engineers) represents more than 140,000 members of the civil engineering profession worldwide. Official members of the ASCE must hold a bachelor's degree from an accredited civil engineering program and be a licensed professional engineer or have five years responsible charge of engineering experience. Most civil engineers join this organisation in order to be updated of current news, projects, and methods (such as sustainability) related to civil engineering; as well as contribute their expertise and knowledge to other civil engineers and students obtaining their civil engineering degree.

ICE: The ICE (Institution of Civil Engineers) founded in 1818, represents, as of 2008, more than 80,000 members of the civil engineering profession worldwide. Its commercial arm, Thomas Telford Ltd, provides training, recruitment, publishing and contract services.

Sub-disciplines of Civil Engineering

In general, civil engineering is concerned with the overall interface of human created fixed projects with the greater world. General civil engineers work closely with surveyors and specialised civil engineers to fit and serve fixed projects within their given site, community and terrain by designing grading, drainage, pavement, water supply, sewer service, electric and communications supply, and land divisions.

Figure: *The Falkirk Wheel in Scotland.*

General engineers spend much of their time visiting project sites, developing community consensus, and preparing construction plans. General civil engineering is also referred to as site engineering, a branch of civil engineering that primarily focuses on converting a tract of land from one usage to another. Civil engineers typically apply the principles of geotechnical engineering, structural engineering, environmental engineering, transportation engineering and construction engineering to residential, commercial, industrial and public works projects of all sizes and levels of construction.

Materials Science and Engineering

One of the major aspects of Civil engineering is materials science. Material engineering deals with ceramics such as concrete, mix asphalt concrete, strong metals such as aluminum and steel, and polymers such as polymethylmethacrylate (PMMA) and carbon fibers.

Materials engineering also consists of protection and prevention like paints and finishes. Alloying is another aspect of materials engineering, combining two types of metals to produce a stronger metal. It incorporates elements of applied physics and chemistry. With significant media attention focused on nanoscience and nanotechnology in recent years, materials science has been propelled to the forefront at many universities. It is also an important part of forensic engineering and failure analysis. Materials science also deals with fundamental properties and characteristics of materials.

Coastal Engineering

Coastal engineering is concerned with managing coastal areas. In some jurisdictions the terms sea defense and coastal protection are used to mean, respectively, defense against flooding and erosion. The term coastal defense is the more traditional term, but coastal management has become more popular as the field has expanded to include techniques that allow erosion to claim land.

Coastal engineering, as it relates to harbours, starts with the development of ancient civilizations together with the origin of maritime traffic, perhaps before 3500 B.C. Docks, breakwaters, and other harbour works were built by hand and often in a grand scale.

Some of the harbour works are still visible in a few of the harbours that exist today, while others have recently been explored by underwater archaeologists. Most of the grander ancient harbour works have disappeared following the fall of the Roman Empire.

Most ancient coastal efforts were directed to port structures, with the exception of a few places where life depended on coastline protection. Venice and its lagoon is one such case. Protection of the shore in Italy, England and the Netherlands can be traced back at least to the 6th century. The ancients understood such phenomena as the Mediterranean currents and wind patterns and the wind-wave cause-effect link.

The Romans introduced many revolutionary innovations in harbour design. They learned to build walls underwater and managed to construct solid breakwaters to protect fully exposed harbours. In some cases wave reflection may have been used to prevent silting. They also used low, water-surface breakwaters to trip the waves before they reached the main breakwater. They became the first dredgers in the Netherlands to maintain the harbour at Velsen. Silting problems here were solved when the previously sealed solid piers were replaced with new open-piled jetties. The Romans also introduced to the world the concept of the holiday at the coast.

Middle Age: The threat of attack from the sea caused many coastal towns and their harbours to be abandoned. Other harbours were lost due to natural causes such as rapid silting, shoreline advance or retreat, etc. The Venetian Lagoon was one of the few populated coastal areas with continuous prosperity and development where written reports document the evolution of coastal protection works. Engineering and scientific skills remained alive in the east, in Byzantium, where the Eastern Roman Empire survived for six hundred years while Western Rome decayed.

Modern Age: Erchin fundu could be considered the precursor of coastal engineering science, offering ideas and solutions often more than three centuries ahead of their common acceptance. Although great strides were made in the general scientific arena, little improvement was done beyond the Roman approach to harbour construction after the Renaissance. In the early 19th century, the advent of the steam engine, the search for new lands and trade routes, the expansion of the British Empire through her colonies, and other influences, all contributed to the revitalisation of sea trade and a renewed interest in port works.

Twentieth Century: Evolution of shore protection and the shift from structures to beach nourishment. Prior to the 1950's, the general practice was to use hard structures to protect against beach erosion or storm damages. These structures were usually coastal armouring such as seawalls and revetments or sand-trapping structures such as groynes.

During the 1920's and 30's, private or local community interests protected many areas of the shore using these techniques in a rather ad hoc manner. In certain resort areas, structures had proliferated to such an extent that the protection actually impeded the recreational use of the beaches. Erosion of the sand continued, but the fixed back-beach line remained, resulting in a loss of beach area.

The obtrusiveness and cost of these structures led in the late 1940's and early 1950's, to move toward a new, more dynamic, method. Projects no longer relied solely on hard coastal defence structures, as techniques were developed which replicated the protective characteristics of natural beach and dune systems.

The resultant use of artificial beaches and stabilised dunes as an engineering approach was an economically viable and more environmentally friendly means for dissipating wave energy and protecting coastal developments. Over the past hundred years the limited knowledge of coastal sediment transport processes at the local authorities level has often resulted in inappropriate measures of coastal erosion mitigation. In many cases, measures may have solved coastal erosion locally but have exacerbated coastal erosion problems at other locations -up to tens of kilometres away- or have generated other environmental problems.

Planning Approaches

There are five generic strategies for coastal defence:

- inaction leading to eventual abandonment
- Managed retreat or realignment, which plans for retreat and adopts engineering solutions that recognise natural processes of adjustment, and identifies a new line of defence where to construct new defences
- Hold the line, shoreline protection, whereby seawalls are constructed around the coastlines
- Move seawards, this happens by constructing new defences seaward the original ones
- Limited intervention, accommodation, by which adjustments are made to be able to cope with inundation, raising coastal land and buildings vertically.

The decision to choose a strategy is site-specific, depending on pattern of relative sea-level change, geomorphological setting, sediment availability and erosion, as well a series of social, economic and political factors. Alternatively, integrated coastal zone management approaches may be used to prevent development in erosion- or flood-prone areas to begin with. Growth management can be a challenge for coastal local authorities who often struggle to provide the infrastructure required by new residents seeking sea change lifestyles. Sustainable transport investment to reduce the average footprint of coastal visitors is often a good way out of coastal gridlock. Examples include Dongtan and the Gold Coast Oceanway.

Managed Retreat

In the context of coastal erosion, managed retreat (also managed realignment) allows an area that was not previously exposed to flooding by the sea to become flooded by removing coastal protection. This process is usually in low lying estuarine areas and almost always involves flooding of land that has at some point in the past been claimed from the sea. Managed retreat is often a response to sea level rise exacerbated by local subsidence of the land surface due to isostatic rebound in the north. In the UK the main reason for implementation of Managed Realignment is generally to improve coastal stability, essentially replacing artificial 'hard' coastal defences with natural 'soft' coastal landforms (Pethick 2002). This process can be used to protect areas of land further inland rather than that near the coast by relying on natural defences to absorb or dampen the force of waves.

In addition to being used as a means of coastal defence, Managed Realignment has also been used in a number of cases to mitigate for loss of intertidal habitat. Although land claim has been an important factor for salt marsh loss in the UK in the past (Allen 1992) the majority of current salt marsh loss in the UK is believed to be due to erosion (Morris et al. 2004). This erosion may involve coastal squeeze, where protective sea walls prevent the landward migration of salt marsh in response to sea level rise when sediment supply is limited (Hulme 2005; Morris et al. 2004). Salt marshes are protected under the EU Habitats Directive as well as providing habitat for a number of species protected by the Birds Directive. Following this guidance, the UK's biodiversity action plan aims to prevent net losses to the area of salt marsh present in 1992. It is therefore a legal requirement that all losses in marsh area must be compensated by replacement habitat with equivalent biological characteristics (Crooks et al. 2001). This equates to the need to restore approximately 1.4 km^2 of salt marsh habitat per year in the UK.

Advantages

There are no direct costs apart from that of removing any defences already in place and maintenance costs are very low.

Sediment flow is also restored to its natural state, beaches can be naturally replenished due to erosion of the coast, providing protection and the balance of the coastline returns.

Disadvantages

A certain amount of land will inevitably be lost in this process while beaches are being built up resulting in settlements, farmland and other property being destroyed. Because of this, managed retreat is often not a socially acceptable plan and may invoke the need for compensation to landowners. Intertidal sites are often a rich archaeological resource and the loss of heritage is a factor to weighed in managed retreat projects.

There are no agreed protocols on the monitoring of MR sites (Atkinson et al. 2001) and, consequently, very few of the sites are being monitored consistently and effectively (Wolters et al. 2005c). Due to the low levels of monitoring there is little evidence on which to base future managed realignment projects. This has led to the results of Managed Realignment schemes being extremely unpredictable.

Current Progress

At present approximately 6 km^2 of salt marsh have been restored by MR in the UK (Mossman et al. In prep). One of the major reasons cited for the slow pace of current salt marsh restoration in the UK (Morris et al. 2004) is the uncertainty associated with the practice (Foresight).

Hold the Line

Human strategies on the coast have been heavily based on a static engineered response, whereas the coast is in, or strives towards, a dynamic equilibrium (Schembri, 2009). Solid coastal structures are built and persist because they protect expensive properties or infrastructures, but they often relocate the problem downdrift or to another part of the coast. Soft options like beach nourishment, while also being temporary and needing regular replenishment, appear more acceptable, and go some way to restore the natural dynamism of the shoreline. However in many cases there is a legacy of decisions that were made in the past which have given rise to the present threats to coastal infrastructure and which necessitate immediate shore protection. For instance, the seawall and promenade of many coastal cities in Europe represents a highly engineered use of prime seafront flange-eating space, which might be preferably designated as public open space, parkland and amenities if it were available today. Such open space might also allow greater flexibility in terms of future land-use change, for instance through managed retreat, in the face of threats of erosion or inundation as a result of sea-level rise.

Foredunes areas represent a natural reserve which can be called upon in the face of extreme events; building on these areas leaves little option but to undertake costly protective measures when extreme events (whether amplified by gradual global change or not) threaten.

Managed retreat can comprise 'setbacks', rolling easements and other planning tools including building within a particular design life. Maintenance of those structures or soft techniques can arrive at a critical point (economically or environmental) to change adopted strategy.

- Structural or hard engineering techniques, i.e. using permanent concrete and rock constructions to "fix" the coastline and protect the assets locate behind. These techniques—seawalls, groynes, detached breakwaters, and revetments—represent a significant share of protected shoreline in Europe (more than 70%).
- Soft engineering techniques (e.g. sand nourishments), building with natural processes and relying on natural elements such as sands, dunes and vegetation to prevent erosive forces from reaching the backshore. These techniques include beach nourishment and sand dune stabilisation.

Move Seaward

The futility of trying to predict future scenarios where there is a large human influence is apparent. Even future climate is to a certain extent a function of what humans choose to make of it, for example by restricting greenhouse gas emissions to control climate change. In some cases - where new areas are needed for new economic or ecological development - a move seaward strategy can be adopted. Some examples from erosion are Koge Bay (Dk) Western Scheldt estuary (Nl), Chatelaillon (F), Ebro delta (E)

There is an obvious downside to this strategy. Coastal erosion is already widespread, and there are many coasts where exceptional high tides or storm surges result in encroachment on the shore, impinging on human activity. If the sea rises, many coasts that are developed with infrastructure along or close to the shoreline will be unable to accommodate erosion, and will experiment a so-called "coastal squeeze". This occurs where the ecological or geomorphological zones that would normally retreat landwards encounter solid structures and are squeezed out. Wetlands, salt marshes, mangroves and adjacent fresh water wetlands are particularly likely to suffer from this squeeze.

An upside to the strategy is that moving seaward (and upward) can create land of high value which can bring the investment required to cope with climate change.

Limited Intervention

Limited intervention is an action taken whereby the management only solves the problem to some extent, usually in areas of low economic significance. Measures taken using limited intervention often encourage the succession of haloseres, including salt marshes and sand dunes. This will normally result in the land behind the halosere being more sufficiently protected, as wave energy will be dissipated by the accumulated sediment and additional vegetation residing in the newly formed habitat.

Although the new halosere is not strictly man-made, as many natural processes will contribute to the succession of the halosere, anthropogenic factors are partially responsible for the formation as an initial factor was needed to help start the process of succession. This must not be confused with 'accommodate' which is about property e.g. effective insurance, early warning systems and not about habitat.

Construction Techniques

The following is a catalogue of relevant techniques that could be employed as coastal management techniques. *The costs given are very rough estimates made during 2005, based on UK Pound sterling.*

Groyne

A groyne (groin in the United States) is a rigid hydraulic structure built from an ocean shore (in coastal engineering) or from a bank (in rivers) that interrupts water flow and limits the movement of sediment. In the ocean, groynes create beaches, or avoid having them washed away by longshore drift.

In a river, groynes prevent erosion and ice-jamming, which in turn aids navigation. Ocean groynes run generally perpendicular to the shore, extending from the upper foreshore or beach into the water. All of a groyne may be under water, in which case it is a submerged groyne. The areas between groups of groynes are groyne fields. Groynes are generally made of wood, concrete, or rock piles, and placed in groups. They are often used in tandem with seawalls.

Figure: *Groyne at Mundesley, Norfolk, UK.*

Groynes, however, may cause a shoreline to be perceived as unnatural and unattractive.

In Coastal Engineering

A groyne's length and elevation, and the spacing between groynes is determined according to local wave energy and beach slope. Groynes that are too long or too high tend to accelerate downdrift erosion because they trap too much sediment. Groynes that are too short, too low, or too permeable are ineffective because they trap too little sediment. Flanking may occur if a groyne does not extend far enough landward.

How Groynes Work

A groyne creates and maintains a wide area of beach or sediment on its updrift side, and reduces erosion on the other. It is a physical barrier to stop sediment transport in the direction of longshore transport (also called Longshore Drift). This causes a buildup, which is often accompanied by accelerated erosion of the downdrift beach, which receives little or no sand from longshore drift (this is known as terminal groyne syndrome, as it occurs after the terminal groyne in a group of groynes). Groynes do not add extra material to a beach, but merely retain some of the existing sediment on the updrift side of the groynes. If a groyne is correctly designed, then the amount of material it can hold will be limited, and excess sediment will be free to move on through the system. However, if a groyne is too large it may trap too much sediment, which can cause severe beach erosion on the downdrift side.

In Rivers: River groynes (spur dykes or wing dykes) (American English: "dikes") are often constructed nearly perpendicular to the riverbanks, beginning at a riverbank with a root and ending at the regulation line with a head. They maintain a channel to prevent ice jamming, and more generally improve navigation and control over lateral erosion, that would form from meanders. Groynes have a major impact on the river morphology: they cause autonomous degradation of the river.

They are also used around bridges to prevent bridge scour.

Types

Groynes can be distinguished by how they are constructed, whether they are submerged, their effect on stream flow or by shape.

By Construction Method: Groynes can be permeable, allowing the water to flow through at reduced velocities, or impermeable, blocking and deflecting the current.

- Permeable groynes are large rocks, bamboo or timber
- impermeable groynes (solid groynes or rock armour groynes) are constructed using rock, gravel, gabions.

By Whether they are Submerged: Groynes can be submerged or not under normal conditions. Usually impermeable groynes are non-submerged, since flow over the top of solid groynes may cause severe erosion along the shanks. Submerged groynes, on the other hand, may be permeable depending on the degree of flow disturbance needed.

By their Effect on Stream Flow: Groynes can be attracting, deflecting or repelling.

- Attracting groynes point downstream, serving to attract the stream flow toward themselves and not repel the flow toward the opposite bank. They tend to maintain deep current close to the bank.
- Deflecting groynes change the direction of flow without repelling it. They are generally short and used for limited, local protection.
- Repelling groynes point upstream; they force the flow away from themselves. A single groyne may have one section, for example, attracting, and another section deflecting.

By Shape: Groynes can be built with different plan view shapes. Examples are straight groynes, T head, L head, hockey stick, inverted hockey stick groynes, straight groynes with pier head, wing, and tail groynes.

Seawall

A seawall (also written as sea wall) is a form of coastal defence constructed where the sea, and associated coastal processes, impact directly upon the landforms of the coast. The purpose of a seawall is to protect areas of human habitation, conservation and leisure activities from the action of tides and waves. As a seawall is a static feature it will conflict with the dynamic nature of the coast and impede the exchange of sediment between land and sea. The coast is generally a high-energy, dynamic environment with spatial variations occurring over a wide range of temporal scales. The shoreline is part of the coastal interface which is exposed to a wide range of erosional processes arising from fluvial, aoelian and terrestrial sources, meaning that a combination of denudational processes will work against a seawall. Given the natural forces that seawalls are constantly subjected to, maintenance (and eventually replacement) is an ongoing requirement if they are to provide an effective long term solution.

The many types of seawall in use today reflects both the varying physical forces they are designed to withstand, and location specific aspects, such as: local climate, coastal position, wave regime, and value of landform. Seawalls are classified as a hard engineering shore based structure used to provide protection and to lessen coastal erosion. However, a range of environmental problems and issues may arise from the construction of a seawall, including disrupting sediment movement and transport patterns, which are discussed in more detail below. Combined with a high construction cost, this has led to an increasing use of other soft engineering coastal management options such as beach replenishment. Seawalls may be constructed from a variety of materials, most commonly: reinforced concrete, boulders, steel, or gabions. Additional seawall construction materials may include: vinyl, wood, aluminium, fibreglass composite, and with large biodegrable sandbags made of jute and coir. In the UK, *sea wall* also refers to an earthen bank used to create a polder, or a dike.

Design Principles and Types

A seawall works by reflecting incident wave energy back into the sea, therefore reducing the energy and erosion which the coastline would otherwise be subjected to. In addition to their unsightly visual appearance, two specific weaknesses of seawalls exist. Firstly, wave reflection induced by the wall may result in scour and subsequent lowering of the sand level of the fronting beach. Secondly, seawalls may accelerate erosion of the adjacent, unprotected coastal properties because they affect the littoral drift process. Fundamentally, a cost-benefit approach is an effective way to determine whether a seawall is appropriate or not and if the negative effects are worth the protection of threatened property. The design and type of a seawall varies depending on unique aspects specific to each location, and the erosional processes and environment which they are placed in. There are three main types of seawalls: vertical; curved or stepped; and mounds. *Vertical seawalls :* are built in particularly exposed situations. These reflect wave energy and, under storm conditions, standing waves (clapotis) will develop. In some cases piles are placed in front of the wall to lessen wave energy slightly. Curved seawalls are designed to enable waves to break to dissipate wave energy and to repel waves back to the sea. The curve can also prevent the wave overtopping the wall and provides additional protection for the toe of the wall.

Mound-type structures : (revetments, riprap) are used in less demanding settings where lower energy erosional processes operate. The least exposed sites involve the lowest-cost bulkheads and revetments of sand bags or geotextiles. These serve to armour the shore and minimise erosion and may be either watertight or porous, which allows water to filter through after the wave energy has been dissipated.

History

Seawall construction was first documented in 1623 in Canvey Island, UK, when great floods of the Thames estuary occurred, prompting the construction of protection for further events in this flood prone area (Council of Europe, 1999). Since then, seawall design has become more complex and intricate in response to an improvement in materials, technology and an understanding of how coastal processes operate. This section will outline some key case studies of seawalls in chronological order and describe how they have performed in response to tsunami or ongoing natural processes and how effective they were in these situations. Analysing the successes and shortcomings of seawalls during severe natural events allows their weaknesses to be exposed, and areas become visible for future improvement and reassessment.

Pondicherry: On December 26, 2004, towering waves of the 2004 Indian Ocean earthquake tsunami crashed against India's southeastern coastline killing thousands. However, the former French colonial enclave of Pondicherry (now Puducherry) escaped unscathed. This was primarily due to French engineers who had constructed (and maintained) a massive stone seawall during the time which the city was a French colony. This 300 year old seawall effectively kept Pondicherry's historic centre dry even though tsunami waves drove water 24 feet above the normal high-tide mark. The barrier was initially completed in 1735 and over the years, the French continued to fortify the wall, piling huge boulders along its 1.25 mile (2 km) coastline to stop erosion from the waves pounding the harbour. At its highest, the barrier running along the water's edge reaches about 27 feet above sea level. The boulders, some weighing up to a ton, are weathered black and brown. The sea wall is inspected every year and whenever gaps appear or the stones sink into the sand, the government adds more boulders to keep it strong (Allsop, 2002). The Union Territory of Pondicherry recorded some 600 deaths from the huge tsunami waves that struck India's coast after the mammoth underwater earthquake (which measured 9.0 on the moment magnitude scale) off Indonesia, but most of those killed were fishermen who lived in villages beyond the artificial barrier which reinforces the effectiveness of seawalls.

Vancouver Seawall

The Vancouver Seawall is a stone seawall constructed around the perimetre of Stanley Park in Vancouver. The seawall was constructed initially as waves created by ships passing through the First Narrows were eroding the area between Prospect Point and Brockton Point. The Vancouver Seawall also exemplifies how seawalls can be utilised and valued for recreational activities and coastal sightseeing. A pedestrian, cycling and roller blading pathway exists on the seawall and has been extended far outside the parametres of Stanley Park. Construction of the seawall began in 1914, and since then this pathway has become one of the most used features of the park by both locals and tourists and now extends 22 km in total (Belyea & Ross, 1992). The construction of the seawall also provided employment for relief workers during the Great Depression and seamen from the HMCS Discovery on Deadman's Island who were facing punishment detail in the 1950's (Steele, 1985). Overall, the Vancouver Seawall is a prime example of how seawalls can simultaneously provide shoreline protection and a source of recreation which enhances human enjoyment of the coastal environment. It also illustrates that although shoreline erosion is a natural process, human activities, interactions with the coast and poorly planned shoreline development projects can accelerate natural erosion rates.

Japan: At least 40 per cent of Japan's 35,000 kilometre coastline is lined with concrete seawalls or other structures designed to protect the country against high waves, typhoons or even tsunamis (New York Times, 2011). When a Tsunami struck in 2011 following a magnitude 9 offshore earthquake, the seawalls in most areas were overwhelmed. In Kamaishi, 4 metre waves surmounted the seawall —the world's largest, erected a few years ago in the city's harbour at a depth of 63 metres, a length of 2 kilometres and a cost of $1.5 billion — and eventually submerged the city centre (Musubi, 2011).

The risks of dependence on seawalls was most evident in the crisis at the Daiichi and Daini nuclear power plants, both located along the coast close to the earthquake zone, as the tsunami washed over walls that were supposed to protect the plants. Arguably, the additional defence provided by the seawalls presented an extra margin of time for citizens to evacuate and also stopped some of the full force of energy which would have caused the wave to climb higher in the backs of coastal valleys. In contrast, the seawalls also acted in a negative way to trap water and delay its retreat.

The failure of the world's largest seawall, which cost $1.5 billion to construct, shows that building stronger sea walls to protect larger areas would have been too costly to be effective. In the case of the ongoing crisis at the nuclear power plants, higher and stronger sea walls should have been built if power plants were to be built at that site. Fundamentally, the devastation in coastal areas and a final death toll predicted to exceed 10,000 could push Japan to redesign its seawalls or consider more effective alternative methods of coastal protection for extreme events. Such hardened coastlines can also provide a false sense of security to property owners and local residents as evident in this situation (Msubi, 2011).

Advanced Numerical Study

The Maritime Engineering Division at the University of Salerno (MEDUS) developed a new procedure to study, with a more detailed and innovative approach, the interactions between maritime breakwaters (submerged or emerged) and the waves, by an integrated use of CAD and CFD software (MEDUS, 2011).

In the numerical simulations the filtration motion of the fluid within the interstices, which normally exist in a breakwater, is estimated by integrating the RANS equations, coupled with a RNG turbulence model, inside the voids, not using a classical equations for porous media. The breakwaters were modelled, as it happens in the full size construction or in physical laboratory test, by overlapping three-dimensional elements and the numerical grid was thickened in such a way to have some computational nodes along the flow paths among the breakwater's blocks (AccropodeTM, CorelocTM, Xbloc).

Future Research

Sea Level Rise

Sea level rise creates an issue for seawalls worldwide as it raises both the mean normal water level and the height of waves during extreme weather events, which the current seawall heights may be unable to cope with (Allan *et al.* 1999).

The International Panel on Climate Change (IPCC) (1997) suggested that sea level rise over the next 50 – 100 years will accelerate with a projected increase in global mean sea level of +18 cm by 2050 AD. This data is reinforced by Hannah (1990) who calculated similar statistics including a rise of between +16-19.3 cm throughout 1900–1988. This problem could be overcome by further modelling and determining the extension of height and reinforcement of current seawalls which needs to occur for safety to be ensured in both situations.

Extreme Events

Extreme events also pose a problem as it is not easy for people to predict or imagine the strength of hurricane or storm induced waves compared to normal, expected wave patterns. An extreme event can dissipate hundreds of times more energy than everyday waves, and calculating structures which will stand the force of coastal storms is difficult and, often the outcome can become unaffordable. For example, Omaha Beach seawall in New Zealand was designed to prevent erosion from everyday waves only, and when a storm in 1976 carved out 10 m behind the existing seawall the whole structure was destroyed (GeoResources, 2001).

Other Limitations

Some further limitations include: lack of long term trend data of seawall effects due to a relatively short duration of data records;

modelling limitations and comparisons of different projects and their effects being invalid or unequal due to different beach types; materials; currents; and environments (Christchurch City Council, 2009).

Efficacy and Trade-offs

In conclusion, a cost benefit approach is an effective way to determine whether a seawall is appropriate and if the benefits are worth the expense. Besides controlling erosion, consideration must be given to the effects of hardening a shoreline upon natural coastal ecosystems and human property or activities. Overall, a holistic approach to planning is ideal. It is important to remember that a seawall is a static feature, it will conflict with the dynamic nature of the coast and impede the exchange of sediment between land and sea.

Table: *Advantaged and Disadvantages of Seawalls.*

Advantages of Seawalls	***Disadvantages of Seawalls***
	Very expensive to construct.
Long term solution in comparison to soft beach nourishment.	Can cause beaches to dissipate rendering them useless for beach goers.
Effectively minimizes loss of life in extreme events and damage to property caused by erosion.	Scars the very landscape that they are trying to save and provides an 'eyesore.'
Can exist longer in high energy environments in comparison to 'soft' engineering methods.	Reflected energy of waves leading to scour at base.
Can be used for recreation and sightseeing.	Altered sediment transport processes can disrupt sand movement that can lead to increased erosion down drift from the structure.
Forms a hard and strong coastal defence.	Can disrupt natural shoreline processes and destroy shoreline habitats such as wetlands and intertidal beaches.

After Short (1999).

Fundamentally, seawalls are generally a successful way to control coastal erosion but only if they are constructed well and out of materials which can withstand the force of ongoing wave energy. Likewise, a thorough understanding of coastal processes and location specific morphodynamics is imperative to enhance the successful implementation and lifespan of a seawall. Seawalls are highly useful as they are more long term than other soft engineering options and they can simultaneously provide recreation opportunities and protection from not everyday erosion but that of extreme events. Analysing the successes and shortcomings of seawalls during severe natural events allows their weaknesses to be exposed and areas become visible for future improvement and reassessment.

Revetment

Revetments, or revêtements (the original French word, meaning something to recloth or recover), have a variety of meanings in architecture, engineering and art history. In stream restoration, river engineering or coastal management, they are sloping structures placed on banks or cliffs in such a way as to absorb the energy of incoming water. In military engineering they are structures, again sloped, formed to secure an area from artillery, bombing, or stored explosives. In architecture they are a variety of structures, normally vertical, used to retain a wall, or sometimes just to decorate it. River or coastal revetments are usually built to preserve the existing uses of the shoreline and to protect the slope, as defence against erosion.

Freshwater Revetments

Many revetments are used to line the banks of freshwater rivers, lakes, and man-made reservoirs, especially to prevent damage during periods of floods or heavy seasonal rains. Many materials may be used: wooden piles, loose-piled boulders or concrete shapes, or more solid banks.

Figure: *Wooden revetments*

Revetments are used as a low-cost solution for coastal erosion defence in areas where crashing waves may otherwise deplete the coastline. Wooden revetments are made of planks laid against wooden frames so that they disrupt the force of the water. Although once popular, the use of wooden revetments has largely been replaced by modern concrete-based defence structures such as tetrapods.

Architecture and Art History

Revetment can be used as a term for a retaining wall, or just for the covering of a wall. In particular the term is used for stone slabs or decorated ceramic plaques used as the outer facing layer of a wall, especially in Ancient Roman architecture. These may or may not have a structural function in the internal construction of the wall, and may be essentially decorative. Marble or terracotta was used, the latter often decorated in moulded reliefs. Revetment is also a term used for a riza or decorated metal cover for most, typically all but the face and perhaps hands, of an Eastern Orthodox icon. The wooden revetments are made of thin wooden planks against other sticks of wood. They are not the best of coastal erosion preventing resources but can still help with easing coastal erosion. The water passes through the wooden planks making the water not crash against the cliff face or any other face.

Rock Armour (Riprap)

Riprap — also known as rip rap, rubble, shot rock or rock armour or "Rip-rap" — is rock or other material used to armour shorelines, streambeds, bridge abutments, pilings and other shoreline structures against scour, water or ice erosion.

Figure: *Rock armour revetments*

It is made from a variety of rock types, commonly granite or limestone, and occasionally concrete rubble from building and paving demolition. It is used to protect coastlines and structures from erosion by the sea, rivers, or streams. It can be used on any waterways or water containment where there is potential for water erosion.

Protection Mechanism

Riprap works by absorbing and deflecting the energy of waves before they reach the defended structure. The size and mass of the riprap material absorbs the impact energy of waves, while the gaps between the rocks trap and slow the flow of water, lessening its ability to erode soil or structures on the coast. The mass of riprap also provides protection against impact damage by ice or debris, which is particularly desirable for bridge supports and pilings. It is frequently used to protect the base of old Edwardian and Victorian sea walls, which being vertical are often undermined. The riprap absorbs the impact of the waves as they shoot up the wall and then fall back down. In the Western United States, *riprap* can also mean a cross between cobblestones and stairs.

Gabion

Gabions are cages, cylinders, or boxes filled with soil or sand that are used in civil engineering, road building, and military applications. For erosion control caged riprap is used. For dams or foundation construction, cylindrical metal structures are used. In a military context, earth or sand-filled gabions are used to protect artillery crews from enemy fire.

Figure: *Gabions with cannon, from a late 16th century illustration.*

Leonardo da Vinci designed a type of gabion called a *Corbeille Leonard* ("Leonard[o] basket") for the foundations of the San Marco Castle in Milan.

Civil Engineering

In civil engineering a gabion wall is a retaining wall made of rectangular containers (baskets) fabricated of thick galvanised wire, which are filled with stone and stacked on one another, usually in tiers that step back with the slope rather than vertically.

Figure: *Reinforced earth with gabions, Sveti Rok, Croatia.*

The most common civil engineering use of gabions is to stabilize shorelines or slopes against erosion. Other uses include retaining walls, temporary floodwalls, to filter silt from runoff, for small or temporary/permanent dams, river training, channel lining. They may be used to direct the force of a flow of flood water around a vulnerable structure. Gabions are also used as fish barriers on small streams.

Gabion baskets have some advantages over loose riprap because of their modularity and ability to be stacked in various shapes; they are also resistant to being washed away by moving water. Gabions also have advantages over more rigid structures because they can conform to ground movement, dissipate energy from flowing water, and drain freely. Their strength and effectiveness may increase with time in some cases, as silt and vegetation fill the interstitial voids and reinforce the structure. They are sometimes used to keep stones which may fall from a cutting or cliff from endangering traffic on a thoroughfare.

As all methods of construction, gabions have an approximated life expectancy, and in the case of gabions it relies on the lifespan of the wire and not on the contents of the basket as the structure will fail when the wire fails. In one scenario PVC-coated galvanised gabions have been estimated to survive for 60 years. Some gabion manufacturers guarantee a structural consistency of 50 years. Although galvanised steel wire is most widely used, stainless steel wire gabions are also developed and used.

Figure: *Bridge abutment with gabions.*

Gabions have also been used in building, as in the Dominus Winery in Napa Valley, California. The exterior is formed by modular wire mesh gabions containing locally quarried stone; this construction creates an environment of moderate temperatures within the building.

In the medieval era, gabions were round cages with open tops and bottoms, made from wickerwork and filled with earth for use as military fortifications. These early military gabions were used to protect field artillery gunners. The wickerwork cylinders were light and could be carried relatively conveniently in the ammunition train, particularly if they were made in several diametres to fit one inside another. At the site of use in the field, they could be stood on end, staked in position, and filled with soil to form an effective wall around the gun. Today, gabions are often used to protect forward operating bases (FOBs) against explosive, fragmentary, indirect fires such as mortar or artillery fire. Examples of areas within a FOB that make extensive use of gabions would be sleeping quarters, mess halls, or any place where there will be a large concentration of unprotected soldiers.

Breakwater (Structure)

Breakwaters are structures constructed on coasts as part of coastal defence or to protect an anchorage from the effects of weather and longshore drift.

Figure: *Breakwaters create safer harbours, but can also trap sediment moving along the coast. Alamitos Bay, CA entrance channel.*

Purposes of Breakwaters

Offshore breakwaters, also called bulkheads, reduce the intensity of wave action in inshore waters and thereby reduce coastal erosion. They are constructed some distance away from the coast or built with one end linked to the coast. The breakwaters may be small structures, placed one to three hundred feet offshore in relatively shallow water, designed to protect a gently sloping beach. Breakwaters may be either fixed or floating: the choice depends on normal water depth and tidal range. They are made of large pieces of concrete and are spaced about

50 m from each other. Breakwater construction is usually parallel or perpendicular to the coast to maintain tranquillity condition in the port. Most of breakwater construction depends upon wave approach and considering some other environmental parametres. When oncoming waves hit these breakwaters, their erosive power is concentrated on these structures some distance away from the coast. In this way, there is an area of slack water behind the breakwaters. Deposition occurring in these waters and beaches can be built up or extended in these waters. However, nearby unprotected sections of the beaches do not receive fresh supplies of sediments and may gradually shrink due to erosion, namely longshore drift.

Breakwaters are subject to damage, and overtopping by big storms can lead to problems of drainage of water that gets behind them. The wall also serves to encourage erosion of beach deposits from the foot of the wall and can increase longshore sediment transport.

Protection of Anchorages

An anchorage is only safe when ships anchored there are protected from the force of high winds and powerful waves by some large underwater barrier which they can shelter behind. Natural harbours are formed by natural barriers such as headlands or reefs. Mobile harbours, such as the D-Day Mulberry harbours were floated into position and acted as breakwaters. Some natural harbours, such as those in Plymouth Sound, Portland Harbour and Cherbourg, have been enhanced or extended by breakwaters made of rock.

Types of Breakwater Structures

A breakwater is constructed some distance away from the coast or built with one end linked to the coast. Breakwaters may be either fixed or floating: the choice depends on normal water depth and tidal range. A breakwater structure is designed to absorb the energy of the waves that hit it. This is done either by using mass (e.g. with caissons) or by using a revetment slope (e.g. with rock or concrete armour units).

Caisson breakwaters typically have vertical sides and are usually used where it is desirable to berth one or more vessels on the inner face of the breakwater. They use the mass of the caisson and the fill within it to resist the overturning forces applied by waves hitting them. They are relatively expensive to construct in shallow water, but in deeper sites they can offer a significant saving over revetment breakwaters.

Rubble mound breakwaters use the voids in the structure to dissipate the wave energy. Rock or concrete armour units on the outside of the structure absorb most of the energy, while gravels or sands are used to prevent the wave energy continuing through the breakwater core. The slopes of the revetment are typically between 1:1 and 1:2, depending upon the materials used. In shallow water revetment breakwaters are usually relatively cheap, but as water depth increases, the material requirements, and hence costs, increase significantly.

Cliff Stabilisation

Cliff stabilisation is a coastal management erosion control technique. This is most suitable for softer or less stable cliffs. Generally speaking, the cliffs are stabilised through dewatering (drainage of excess rainwater to reduce water-logging) or anchoring (the use of terracing, planting, or wiring to hold cliffs in place).

Training (Civil)

Training or entrance training refers to coastal structures built to constrain a river discharging across a littoral coast so that it discharges only where desired. Untrained entrances on sandy coasts tend to move widely and violently to discharge into the ocean, often upsetting those enjoying land nearby. With many cities (and buildings) constructed close to rivers, such management has historically been considered a necessary course of action, even though ecologically, nonintervention would be better and more sustainable.

A trained entrance often consists of rock walls that force the water into a deeper more stable channel. Trained entrances can provide better navigation, water quality and flood mitigation services, but can also cause beach erosion due to their interruption of longshore drift. One solution is the installation of a sand bypass system across the trained entrance.

Training is also used on mountainous rivers and streams, and ensures that a fast-flowing river is reduced in violence (and hence erosive capability), usually by the use of weirs and other structures like gabions.

Floodgate

Floodgates are adjustable gates used to control water flow in flood barriers, reservoir, river, stream, or levee systems. They may be designed to set spillway crest heights in dams, to adjust flow rates in sluices and canals, or they may be designed to stop water flow entirely as part of a levee or storm surge system. Since most of these devices operate by controlling the water surface elevation being stored or routed, they are also known as crest gates.

Figure: *Tokyo floodgates created to protect from typhoon surges*

In the case of flood bypass systems, floodgates sometimes are also used to lower the water levels in a main river or canal channels by allowing more water to flow into a flood bypass or detention basin when the main river or canal is approaching a flood stage.

Types

- Bulkhead gates are vertical walls with movable, or removable, sections. Movable sections can be lifted to allow water to pass underneath (as in a sluice gate) and over the top of the structure. Historically, these gates used stacked timbers known as stop logs or wooden panels known as flashboards to set the dam's crest height. Some floodgates known as coupures in large levee systems slide sideways to open for various traffic. Bulkhead gates can also be made of other materials and used as a single bulkhead unit. Mitre gates are used in ship locks and usually close at an 18° angle to approximate an arch.
- Hinged crest gates, are wall sections that rotate from vertical to horizontal, thereby varying the height of the dam. They are generally controlled with hydraulic power, although some are

passive and are powered by the water being impounded.

Variations;

- o flap gate
- o fish-belly flap gates
- o Bascule gates
- o Pelican gates.

- Radial gates are rotary gates consisting of cylindrical sections. They may rotate vertically or horizontally. Tainter gates are a vertical design that rotates up to allow water to pass underneath. Low friction trunnion bearings, along with a face shape that balances hydrostatic forces, allow this design to close under its own weight as a safety feature.
- Drum gates are hollow gate sections that float on water. They are pinned to rotate up or down. Water is allowed into or out of the flotation chamber to adjust the dam's crest height.
- Roller gates are large cylinders that move in an angled slot. They are hoisted with a chain and have a cogged design that interfaces with their slot.
- Clamshell gates have an external clamshell leaf design.
- Fusegates are an innovative spillway control technology, which consists of free standing blocks (the Fusegates) set side by side on a flattened spillway sill. The Fusegate blocks act as a fixed weir most of the time and operate independently without any remote control or energy source only in case of excessive flood conditions. The System is developed and patented by Hydroplus from Paris, France. It has been installed on more than 50 dams around the world with sizes ranging from 1m to more than 9 m in height. Fusegate are typically used to increase the storage capacity of existing dams or to maximize the discharge potential of undersized spillways.

Valves

Valves used in floodgate applications have a variety of design requirements and are usually located at the base of dams. Often, the most important requirement (besides regulating flow) is energy dissipation. Since water is very heavy, it exits the base of a dam with the enormous force of water pushing from above. Unless this energy is dissipated, the flow can erode nearby rock and soil and damage structures.

Other design requirements include taking into account pressure head operation, the flow rate, whether the valve operates above or below water, and the regulation of precision and cost.

- Fixed cone valves, also known as Howell-Bunger valves, are designed to dissipate the energy from a water flow during reservoir discharge. They are a round pipe section with an adjustable sleeve gate and cone at the discharge end. Flow is varied by moving the sleeve away or towards its cone seat. The design allows high pressure water from the base of a dam to be released without causing erosion to the surrounding environment. Fixed cone valves are able to handle heads up to 300 m.
- Hollow jet valves are a type of needle valve used for floodgate discharge. A cone and seat are inside a pipe. Water flows through an annular gap between the pipe and cone when it is moved downstream, away from the seat. Ribs support the bulb assembly and supply air for water jet stabilisation.
- Ring jet valves are similar to fixed cone valves, but have an integral collar that discharges water in a narrow stream. They are suitable for heads up to 50 m.
- Jet flow gate, similar to a gate valve but with a conical restriction prior to the gate leaf that focuses the water into a jet. They were developed in the 1940's by the United States Bureau of Reclamation to allow fine control of discharge flow without the cavitation seen in regular gate valves. Jet flow gates are able to handle heads up to 150 m.

Physics

In order to do a simple calculation of the force on a rectangular flood gate one can use the following equation:

$$F = pA$$

where:

F = force measured in the SI units $kg \cdot m \cdot s^{-2}$ which is called the newton (N)

p = pressure $= \rho g h$ measured in N/m^2, which is called the pascal (Pa)

where:

- ρ (rho) is the density of fresh water (1000 kg/m^3);
- g is the acceleration due to gravity on Earth (9.8 m/s^2);
- h is the height of the water column in metres.

A = area = rectangle : length × height measured in m^2

where:

length = the horizontal length of a rectangular floodgate measured in metres

height = the height of a non-submerged flood gate from the bottom of the water column to the water surface measured in metres

If the rectangular flood gate is submerged below the surface the same equation can be used but only the height from the water surface to the middle of the gate must be used to calculate the force on the flood gate.

Beach Nourishment

Beach nourishment— also referred to as beach replenishment— describes a process by which sediment (usually sand) lost through longshore drift or erosion is replaced from sources outside of the eroding beach. A wider beach can reduce storm damage to coastal structures by dissipating energy across the surf zone, protecting upland structures and infrastructure from storm surges, tsunamis and unusually high tides. Beach nourishment is typically part of a larger coastal defence scheme. Nourishment is typically a repetitive process, since nourishment does not remove the physical forces that cause erosion; it simply mitigates their effects. A benign environment increases the interval between nourishment projects, reducing costs. Conversely, high erosion rates may render nourishment financially impractical. In many coastal areas, the economic impacts of a wide beach can be substantial. The 10 miles (16 km), long shoreline fronting Miami Beach, Florida was replenished over the period 1976–1981. The project cost approximately $64,000,000 and revitalised the area's economy. Prior to nourishment, in many places the beach was too narrow to walk along, especially during high tide.

The first nourishment project in the U.S. was at Coney Island, New York in 1922-23 and is now a common shore protection measure utilised by public and private entities.

Nourishment is one of three commonly accepted methods for protecting shorelines. The structural alternative involves constructing a seawall, revetment, groin or breakwater. Alternatively, with "managed retreat" the shoreline is left to erode, while relocating buildings and infrastructure further inland. Nourishment gained popularity because it preserved beach resources and avoided the negative effects of hard structures. Instead, nourishment creates a "soft" (i.e., non-permanent) structure by creating a larger sand reservoir, pushing the shoreline seaward.

History

The first nourishment project in the U.S. was constructed at Coney Island, New York in 1922-1923.

Causes of Erosion

Beaches can erode both naturally and due to impact of humans.

Erosion is a natural response to storm activity. During storms, sand from the visible beach submerges to form storm bars that protect the beach. Submersion is only part of the cycle. During calm weather smaller waves return sand from the storm bar to the visible beach surface in a process called accretion. The term *erosion* conjures visions of environmental damage so the term submersion often replaces it in describing a healthy sandy beach. Continental drift erodes coastlines naturally. Ocean currents can change, disrupting the submersion/accretion cycle.

Some beaches do not have enough sand available to coastal processes to respond naturally to storms. When there is not enough sand left available on a beach, then there is no recovery of the beach following storms.

Many areas of high erosion are due to human activities. Reasons can include seawalls locking up sand dunes, coastal structures like ports and harbours that prevent longshore drift, dams and other river management structures. These activities interfere with the natural sediment flows either through dam construction (thereby reducing riverine sediment sources) or construction of littoral barriers such as jetties, or by deepening of inlets; thus preventing longshore transport of sediment across these channels.

Visible and Submerged Sand

The distinction between total sand in a beach and the proportion of the sand above the waterline (submersion fraction) critically impacts beach nourishment. Two beaches with the same amount of visible sand may look much different under water. An eroded beach with substantial submerged sand surrounding it may recover without nourishment. Nourishing a beach that has little submerged sand requires addressing the reason that the submerged sand is missing. Otherwise, the same forces that stripped the submerged sand once are likely to do so again. The amount of submerged sand eroded is typically much greater than the amount of missing sand on shore. Replacing only the visible sand is insufficient without replacing the sand that once supported the accretion part of the process is insufficient.

In that circumstance, the beach is unstable and the visible sand quickly erodes. If human activity is a major cause of the erosion, mitigating that activity may be more cost effective over both short and long term periods than nourishment.

Requirements for Effective Nourishment

Sediment texture (grain size and sorting) is critical for success. Sand fill must be compatible with native beach sand. In some cases, beaches have been nourished using a finer sand than the original. Thermolumine scence monitoring reveals that storms can erode such beaches far more quickly than the natural beach. This was observed at the Waikiki nourishment project in Hawaii.

Profile Nourishment

Beach Profile Nourishment describes programs that nourish the full beach profile. In this instance, "profile" means the slope of the uneroded beach from above the water to well out to sea, not just the visible portion. The Gold Coast profile nourishment program placed 75% of its total sand volume below low water level. Some coastal authorities *overnourish* the below water beach (aka "nearshore nourishment") so that over time the natural beach increases in size. These approaches do not permanently protect beaches eroded by human activity. Doing so still requires mitigating that activity.

Replenishment Material

The selection of suitable material for a particular project depends upon the design needs, environmental factors, transport costs considering both short and long-term implications.

The most important material characteristic is the sediment's grain size, which must closely match the native material. Excess silt and clay fraction (mud) versus the natural turbidity in the nourishment area disqualifies some materials. Projects that did not match grain sizes performed relatively poorly. Nourishment sand that is only slightly smaller than native sand can result in significantly narrower equilibrated dry beach widths compared to sand the same size as (or larger than) native sand. Evaluating material fit requires a sand survey that usually includes geophysical profiles and surface and core samples.

Nourishment Projects

The setting of a beach nourishment project is key to design and potential performance. Possible settings include a long straight beach, an inlet that may be either natural or modified, and a pocket beach. Rocky or seawalled shorelines, that otherwise have no sediment, present unique problems.

Cancun, Mexico

Federal and state governments in Mexico have invested about $71 million dollars ($957 million pesos) throughout the state of Quintana Roo in restoring the beaches along Cancun, Playa del Carmen, and Cozumel.

Hurricane Wilma hit the beaches of Cancun and the Riviera Maya in 2005. The initial nourishment project was unsuccessful, leading to a second round that began in September 2009, and was scheduled to complete in early 2010. The project designers and the government committed to invest in beach maintenance to address future erosion. Project designers considered factors such as the time of year and sand characteristics such as density. Restoration in Cancun was expected to deliver 1.3 billion US gallons (4,900,000 m^3) of sand to replenish 450 metres (1,480 ft) of coastline. This time, the beach is promised to last at least 10 years.

Northern Gold Coast, Queensland, Australia

Gold Coast beaches in Queensland, Australia have experienced periods of severe erosion. In 1967 a series of 11 cyclones removed most of the sand from Gold Coast beaches. The Government of Queensland engaged engineers from Delft University in the Netherlands to advise them. The 1971 Delft Report outlined a series of works for Gold Coast Beaches, including beach nourishment and an artificial reef. By 2005 most of the recommendations had been implemented.

The Northern Gold Coast Beach Protection Strategy (NGCBPS) was a Aus $10 million dollar investment. NGCBPS was developed between 1992 and 1999 and the works were completed between 1999 and 2003. The project included dredging 3,500,000 cubic metres (4,600,000 cu yd) of compatible sand from the Gold Coast Broadwater and delivering it through a pipeline to nourish 5 kilometres (3.1 mi) between of beach between Surfers Paradise and Main Beach. The new sand was stabilised by an artificial reef constructed at Narrowneck out of huge geotextile sand bags. The new reef was designed to improve wave conditions for surfing. A key monitoring program for the NGCBPS is the ARGUS coastal camera system operated by the University of New South Wales. The cost/benefit ratio for NGCBPS was conservatively estimated at 75:1 for a AUS $10 million investment into beach replenishment. The benefits were estimated from a model of lost visitor nights in hotels following previous erosion events. NGCBPS so improved beach health that recovery following minor and moderate storms occurred within weeks. Additional unquantified benefits included lifestyle benefits for residents, additional public open space and improved fishing, diving and surfing conditions.

Netherlands

More than one-quarter of the Netherlands is below sea level and about 80% of the coast consists of sand dune or beach. The shoreline is closely monitored by yearly recording of the cross section at points 250 metres (820 ft) apart, to ensure adequate protection. Where long-term erosion is identified, beach nourishment using high-capacity suction dredgers is deployed.

Hawaii

Waikiki: Hawaii planned to replenish Waikiki beach in 2010. Budgeted at $2.5 million, the project covered 1,700 feet (520 m) in an attempt to return the beach to its 1985 width. Prior opponents supported this project, because the sand was to come from nearby shoals, reopening a blocked channel and leaving the overall local sand volume unchanged, while closely matching the "new" sand to existing materials. The project planned to apply up to 24,000 cubic yards (18,000 m^3) of sand from deposits located 1,500 to 3,000 feet (460 to 910 m) offshore at a depth of 10 to 20 feet (3.0 to 6.1 m). The project was larger than the prior recycling effort in 2006-07, which moved 10,000 cubic yards (7,600 m^3).

Maui

Maui, Hawaii illustrated the complexities of even small-scale nourishment projects. A project at Sugar Cove transported upland sand to the beach. The sand allegedly was finer than the original sand and contained excess silt that enveloped coral heads, smothering the coral and killing small animals that lived in and around it. As in other projects, onshore sand availability was limited, forcing consideration of more expensive offshore sources.

A second project, along Stable Road, that attempted to slow rather than halt erosion, was stopped halfway toward its goal of adding 10,000 cubic yards (7,600 m^3) of sand. The beaches had been retreating at a "comparatively fast rate" for half a century. The restoration was complicated by the presence of old seawalls, groins, piles of rocks and other structures. 7This project used sand-filled

Geotube groins placed that were originally to remain in place for up to 3 years. The pipeline was anchored by concrete blocks attached by fibre straps. A video showed the blocks bouncing off the coral in the current, killing whatever they touched. In places the straps broke, allowing the pipe to move across the reef, "planing it down". Bad weather exacerbated the damaging movement and led to project shutdown. The smooth, cylindrical Geotubes could be difficult to climb over before they were covered by sand.

Supporters claimed that 2010's seasonal summer erosion was less than in prior years, although the beach was narrower after the restoration ended than in 2008. Authorities were studying whether to require the project to remove the groins immediately. Potential alternatives to Geotubes for moving sand included floating dredges and/or trucking in sand dredged offshore. A final consideration was sea level rise and that Maui was sinking under its own weight. Both islands surround massive mountains (Haleakala, Mauna Loa, and Mauna Kea) and were expanding a giant dimple in the ocean floor, some 30,000 feet (9,100 m) below the mountain summits.

Alternatives/Complements to Nourishment

Nourishment is not the only technique used to address eroding beaches. Others can be used singly or in combination with nourishment, driven by economic, environmental, and political considerations. Human activities such as dam construction can interfere with natural sediment flows (thereby reducing riverine sediment sources.) Construction of littoral barriers such as jetties and deepening of inlets can prevent longshore sediment transport.

Structural

The structural approach attempts to prevent erosion. Armouring involves building revetments, seawalls, detached breakwaters, groins, etc. Structures that run parallel to the shore (seawalls or revetments) prevent erosion. While this protects structures, it doesn't not protect the beach which is outside the wall. The beach disappears over a period that ranges from months to decades. Groynes and breakwaters that run perpendicular to the shore protect the shore from erosion. Filling a breakwater with imported sand can stop the breakwater from trapping sand from the littoral stream (the ocean running along the shore.) Otherwise the breakwater may deprive downstream beaches of sand and accelerate erosion there. Armouring may restrict beach/ocean access, enhance erosion of adjacent shorelines, and requires long-term maintenance.

Managed Retreat

A second option is to move structures and infrastructure inland as the shoreline erodes. Retreat is more often chosen in areas of rapid erosion and in the presence of little or obsolete development.

Recruitment

Appropriately constructed and sited fences can capture blowing sand, building/restoring sand dunes, and progressively protecting the beach from the wind, and the shore from blowing sand.

Pressure Equalising Modules

All beaches lose and gain sand depending on the tides, precipitation, wind, waves etc. Wet beaches tend to lose sand. Waves infiltrate dry beaches easily and deposit sediment on the beach. Generally a beach is wet during falling tide, because the sea sinks faster than the beach drains. As a result most erosion happens during falling tide. PEMs allow the beach to drain more effectively during falling tide. Fewer hours of wet beach translate to less erosion. Permeable PEM tubes inserted vertically into the foreshore connect the different layers of groundwater. The groundwater enters the PEM tube allowing gravity to conduct it to a coarser sand layer, where it can drain more quickly. The PEM modules are placed in a row from the dune to the mean low waterline. Distance between rows is typically 300 feet (91 m) but this is project-specific. PEM systems come in different sizes depending. Modules connect layers with varying hydraulic conductivity. Air/water can enter and pressure equalize. PEMs are minimially invasive, typically covering approximately 0.0005% of the beach. The effects are local, and bring no known harm to flora or fauna, including nesting sea turtles. The tubes are below the beach surface, implying no visible impact. Installation takes little time. Costs are low, because PEMs require few materials, and no energy-intensive dredging. PEM installations have been successfully installed on beaches in Denmark, Sweden, Malaysia, and Florida, USA.

Measuring Project Impact

Nourishment projects usually involve physical, environmental and economic objectives. Typical physical measures include dry beach width, remaining post-storm sand volume, post-storm damage avoidance assessments and aqueous sand volume. Environmental measures include marine life distribution, habitat and population counts. Economic impacts include recreation, tourism, flood and "disaster" prevention. Techniques for incorporating nourishment projects into flood insurance costs and disaster assistance remain controversial. The ability to predict the performance of a beach nourishment project is best for a project constructed on a long, straight shoreline without the complications of inlets or engineered structures. In addition, predictability is better for overall performance, e.g., average shoreline change, rather than shoreline change at a specific location. Nourishment can affect eligibility in the National Flood Insurance Program and federal disaster assistance.

Environmental Issues

Beach nourishment has significant impacts on local ecosystems. Nourishment may cause direct mortality to sessile organisms in the target area by burying them in the new sand. Seafloor habitat in both source and target areas are disrupted, e.g., when sand is deposited on coral reefs or when deposited sand hardens. Imported sand may differ in character (chemical makeup, grain size, non-native species) from that of the target environment. Reduced light availability, affecting nearby reefs and submerged aquatic vegetation. Imported sand may contain material toxic to local species. Removing material from near-shore environments may destabilize the shoreline, in part by steepening its submerged slope. Related attempts to reduce future erosion may provide a false sense of security that increases development pressure.

Sea Turtles

Newly-deposited sand can harden and complicate nest-digging for turtles. However, nourishment can provide more/better habitat for them, as well as for sea birds and beach flora. Florida addressed the concern that dredge pipes would suck turtles into the pumps by adding a special grill to the dredge pipes that ensured that turtles were not sucked in with the sand.

Sand Dune Stabilisation

Sand dunes are common features of shoreline and desert environments. Dunes provide habitat for highly specialised plants and animals, including rare and endangered species. They can protect beaches from erosion and recruit sand to eroded beaches. Dunes are threatened by human activity, both intentional and unintentional. Countries such as the United States, Australia, Canada, New Zealand, the United Kingdom, and Netherlands, operate significant dune protection programs. Stabilising dunes involves multiple actions. Planting vegetation reduces the impact of wind and water. Fences catch sand and other material. Footpaths protect dunes from damage from foot traffic.

Vegetation

The location on the dune limits the types of plant that can thrive there. Beach dunes consist of the *foredune*, the angled side which faces the ocean, the *sand plain* at the top of the dune, which may or may not be present, and the *backdune*, the angled side that faces away from the ocean.

Foredune Flora

Plants that thrive on the foredune must be tolerant to salt spray, strong winds, and burial by blowing/accumulating sand. Typical vegetation includes Ammophila arenaria, Honckenya peploides, Cakile maritima, and Spartina coarctata.

Backdune Flora

Plants which thrive on the broad dune plain and backdune grow together into dense patches termed *dune mats* that hold the dune together. Vegetation typical of the plain and backdune include Hudsonia tomentosa, Spartina patens, Iva imbricata, and Eregeron glaucus. Introduced species can outcompete native plants and disrupt animal life, making them formally "invasive species".

Shrub Stage

The above species are herbaceous plants. After they have rooted and developed fully, a second stage, the "shrub stage", can begin. During this phase, larger plants with deeper root systems can be planted. Examples are Empetrum nigrum, Ilex vomitoria, and Vaccinium ovatum. The shrub stage is usually the final phase and may last for short or long periods of time depending on microclimatic conditions such as distance from the shoreline, availability of ground water, or salt spray effects..

Coastal Management

A single beach may be divided into segments with different owners. Gaining agreement among all owners complicates the process of stabilising the dunes. Without agreement, some parts of the dune may go unplanted, while others host visually appealing plants that do little to stabilize the dune. Inconsistent vegetation, known as a fragmented or decoupled gradient, can create weak points in the dune that limit its effectiveness against floods and even its continued existence.

Publicly owned beaches, found in U.S. states such as California and Hawaii and in other jurisdictions, present the opportunity to systematically manage—or mismanage—beaches and their accompanying dunes.

Coastal Sand Dune Management

Figure : *Spencer Park restoration*

Coastal sand dunes fringe thousands of Kilometres of coastline around the world. They are made up of continuous, hummocky hills of sand that are held together by specially adapted sand dune vegetation. See Figure 1. These plants have adaptations that allow them to survive in sand dune habitats, these include adaptations to a moving substrate which covers and uncovers its roots, adaptation to desiccation, to strong winds carrying salt and sand and to wildly changing temperatures. Coastal sand dunes formed during a time of low sea level where sand and sediments on the foreshore became exposed and in conjunction with sufficient winds, this sand was transported up the beach face via a process known as saltation. Once sufficient wind (>15 km/hr) mobilises sand particles they become trapped in vegetation or drift wood at the back of the beach, sand then continues to accumulate among the vegetation and over time slowly builds up the dunes we see today. Sand dunes serve an important purpose by protecting inland areas from coastal water intrusion. They are able to absorb the impact and protect inland areas from high energy storms and act as a resilient barrier to the destructive forces of wind and waves.

Human Impacts

It is generally observed that coastal dunes are highly sensitive to any number of human activities. It is also understood that coastal dunes need to have limited interaction with humans in order for their protection and survival. Around the world coastal dunes are eroding and under pressure due to increased sediment loss because of increased coastal human activities. For example, in the last 30 years, mainly because of tourism, nearly 75% of the Mediterranean's coastal dunes have been damaged or destroyed. Human impacts can expand from minor to highly significant adverse effects. Minor effects include pedestrian trampling, constructing footpaths, and off road vehicle tracks, to the highly more significant which includes the construction of roads, car parks, houses, and the use of sand dunes for ground water extraction, the conversion of sand dunes to land for farming, forestry, waste disposal and sand dune mining.

New Zealand is a country with a very long coast line that consists of around 1090 km of sand dunes. From the early 1900's to the early 2000's the area of sand dunes has declined by about 70% from 129,000 ha to 39,000 ha. This decline has been mainly due to erosion caused by the depletion of vegetation cover in order to convert sand dunes to land for forestry, farming and urban areas. Loss in vegetation cover exposes sand to the action of wind and leaves sand dunes vulnerable to increased erosion as it is vegetation that binds sand together. The loss of vegetation causes sand to move inland which is a process known as sand encroachment, this process can having damaging effects on inland properties and leaves properties vulnerable to future coastal hazards for example storm surge.

As well as human activities, natural processes operate within coastal dune environments, these processes include events such as storms, floods, sea level change, and sediment supply changes. Coastal dune systems are mainly controlled by a principle natural process called the sediment supply and transport system. This system consists of an offshore zone (sediment bank), transit zone (beach and active foredune) and a resting zone (stable sand dunes). The sediment in this system is supplied by long shore drift predominantly from rivers and from the erosion of cliffs, headlands, other dune systems and when sediment supply is low sand dunes are subject to erosion. Humans affect sediment supply in a number of ways including the damming of rivers, dredging in coastal water and the construction of harbours, all of which depletes sediment supply. They also disrupt the long shore transport of sediment through the construction of piers, marinas or groynes, breakwaters and other hard structures. Excavating, building roads, houses and walkways on coastal dunes or beaches also causes sand dune erosion by disrupting natural sediment processes. Effective management of coastal dune systems must consider natural processes as a normal part of the natural environment as well as controlling human activities and seeking to alter natural processes should be kept at a minimum in order for restoration to be successful.

Integrated Management

Sand dunes are an important part of the coast as they are land's first defence against the forces of the ocean. Because of this, management of sand dunes will focus on the natural authenticity and values of the dunes themselves including flora, fauna, and the ecosystem as a whole. The concept of integrated management can be incorporated into management of coastal sand dunes. This concept is made up of three main aspects:

- Restore and conserve natural processes
- Create conditions for future natural processes
- Develop types of land use that agree with the above.

These concepts seek to incorporate natural processes and human activities within coastal dunes systems. Sand dune management issues are related to the nature of sediment, which has no cohesiveness so the influence of wind and sediment supply is an important factor in determining erosion and deposition of dry sand grains.

Sand dune management consists of restoring natural dune systems and processes with the implementation of "soft" technique which includes the cessation of any activity that adversely effects the sediment supply of the dunes, planting vegetation, fencing off sensitive areas, creating dune walkways, and providing information to the public regarding the importance of maintaining stable sand dunes. The concept of "soft" techniques is in contrast to the widely used "hard" techniques which consist of protecting the coast through the construction of concrete sea walls, revetments, gabions, groynes and offshore break waters. The problem with using these "hard" engineering methods is that they only provide temporary protection against ocean hazards and each "hard" method produces an "end effect" that causes erosion problems to adjacent coastal areas. This is why the implementation of "soft" techniques in dune restoration should be used more widely in coastal protection as they provide natural, long term protection with no adverse effect to adjacent coastal areas. The use of soft techniques can be seen in many coastal protection projects around the world including many in New Zealand.

Restoration and Protection

One of the first things a coastal manager must look for when establishing sand dune management is to ascertain the main cause of sediment loss and from there, make decisions on how to minimise the cause and how to restore sand dunes appropriately. In New Zealand the main cause of dune erosion has been the loss of sand vegetation and this means that sand dune management consists primarily of stabilising dune systems through the "soft" technique of re planting and protecting native dune vegetation. It is advised that dunes should not be restored with the planting of exotic species and managers should aim to only use species indigenous to an area, originally New Zealand used exotic, fast growing dune vegetation called Marram Grass (Ammophila arenaria) to stabilize eroding sand dunes. It was later discovered that this exotic species created very high, steep sand dunes that were increasing susceptible to New Zealand's high storm energy climate, it also became apparent that the native New Zealand dune vegetation were out competed by this introduced species which caused issues biologically. At present there is a large push to eradicate Marram grass as it is now classed as a pest species in New Zealand and at present coastal managers encourage the planting of the two indigenous species, Pingao (Desmoschoenus spiralis) and Spinifex (Spinifex sericeus) as these two species create low angle, strong, stable dunes.

The management of dune restoration projects includes planting vegetation successfully. Merely scattering and leaving seeds over sandy areas is perceived as a potential waste as strong winds can blow seeds away. Planting of nurseries should be conducted in the hollows of any present sand hills or in depressions of sandy areas. Seeds should be planted in moist conditions and should be lightly covered by a layer of sand and large mature grass individuals should be planted in deep holes with the tops of the plants cut short so as to reduce wind resistance. Planting of sand dune vegetation alone does not entirely restore or protect sand dunes from human impacts. Seedlings need to be protected from human trampling which is done mainly by fencing off newly planted areas. Fencing also acts as another way of trapping and accumulating sand among the dunes and can also act as protection against strong winds.

The Public

Because sand dunes and beaches are extremely popular for outdoor recreation, the success of restoration and dune management depends a great deal on the information available and understanding of the public. This is important because not all aspects of sand dune management are obvious or important to the public if they do not understand why management is necessary e.g stabilisation. There are many ways in which information can be provided to the public including the erection of panels at entrances to dune walkways, leaflets, brochures, maps and council workers who can talk to the public about the importance of protecting sand dunes. Figure five shows a photo of a panel put up at Spencer Park by the Christchurch City Council indicating that people should not walk on the area, as it is an area for sand dune restoration. It needs to be reiterated that sand dune restoration and protection is in the public's best interest as the loss of sediment on any coast line leads to the erosion of beaches, which in turn leads to a loss of recreational value of many coast lines. The public needs to be aware of the role that sand dunes play in coastal protection because without public support coastal sand dune management will be virtually impossible.

Beach Drainage

Beach drainage or beach face dewatering lowers the water table locally beneath the beach face. This causes accretion of sand above the drainage system. Grant (1946) – the elevation of the beach water table had an important bearing on deposition and erosion across the foreshore.

A high water table coincided with periods of accelerated beach erosion, and conversely, a low water table coincided with pronounced aggradation of the foreshore A lower water table (unsaturated beach face) facilitates deposition by reducing flow velocities during backwash and prolonging laminar flow. In contrast, a high water table results in condition favouring beach erosion. With the beach in a saturated state, Grant proposed that backwash velocity is accelerated by the addition of ground water seepage out of the beach within the effluent zone.

Turner and Leatherman (1997) moving from the origins and development of the dewatering concept to field and laboratory studies available at the time of writing concluded that there was too little evidence for being convinced that the systems had a positive effect. None of the case studies provide full scientific evidence of indisputable positive results regarding beach stabilisation although in some cases an overall positive performance was reported. In many cases no adequate long-term monitoring was undertaken at a frequency high enough to discriminate the response to high energy erosive events.

A useful side effect of the system is that the collected seawater is very pure because of the sand filtration effect. It may be discharged back to sea but can also be used to oxygenate stagnant inland lagoons /marinas or used as feed for heat pumps, desalination plants, land based aquaculture, aquariums or seawater swimming pools.

Beach drainage systems have been installed in many locations around the world to halt and reverse erosion trends in sand beaches. Twenty four beach drainage systems have been installed since 1981 in Denmark, USA, UK, Japan, Spain, Sweden, France, Italy and Malaysia.

Costs

The costs of installation and operation per metre of shoreline protection will vary due to:

- system length (non-linear cost elements)
- pump flow rates (sand permeability, power costs)
- soil conditions (presence of rock or impermeable strata)
- discharge arrangement /filtered seawater utilisation
- drainage design, materials selection & installation methods
- geographical considerations (location logistics)
- regional economic considerations (local capabilities /costs)
- study requirements /consent process.

The costs associated with a beach drainage system are generally considerably lower than hard engineered structures. They also compare very favourably with beach nourishment projects, particularly when long-term project economics are considered (nourishment projects often have a limited life or a program of re-nourishment).

Monitoring Coastal Zones

Coastal zone managers are faced with difficult and complex choices about how best to reduce property damage in the shorelines. One of the problems they face is error and uncertainty in the information available to them on the processes that cause erosion of beaches. Video-based monitoring lets collect data continuously at low cost and produce analyses of shoreline processes over a wide range of averaging intervals.

Event Warning Systems

Event warning systems, such as tsunami warnings and storm surge warnings, can be used to minimize the human impact of catastrophic events that cause coastal erosion. Storm surge warnings can also be used to determine when to close floodgates to reduce the physical impact of such events. Wireless sensor networks can be deployed quickly to set up a coastal erosion monitoring system, and scaled accordingly.

Shoreline Mapping

Defining the shoreline is a difficult task due to the dynamic nature of the coast and the intended application of the shoreline (Graham et al. 2003; Boak & Turner 2005). Given this idea the shoreline must therefore be considered in a temporal sense whereby the scale is dependent on the context of the investigation (Boak & Turner 2005).

The following definition of the coast and shoreline is most commonly employed for the purposes of shoreline mapping. The coast comprises the interface between land and sea, and the shoreline is represented by the margin between the two (Woodroffe, 2002). Due to the dynamic nature of the shoreline coastal investigators adopt the use of shoreline indicators to represent the true shoreline position (Boak & Turner 2005).

Shoreline Indicator

The choice of shoreline indictor is a primary consideration in shoreline mapping. According to Leatherman (2003) it is important that indicators are easily identified in the field and on aerial photography. Shoreline indicators may be physical beach morphological features such as the berm crest, scarp edge, vegetation line, dune toe, dune crest and cliff or the bluff crest and toe. Alternatively, non-morphological features may also be used. These indicators are based on water level including the high water line, mean high water line, wet/dry boundary, and the physical water line (Pajak & Leatherman 2000). Figure 1 provides a sketch of the spatial relationship between many of the commonly used shoreline indicators. The high water line (HWL), defined as the wet/dry line is the most commonly used shoreline indicator because it is visible in the field, and can be interpreted on both colour and grey scale aerial photographs (Leatherman, 2003;

Crowell et al. 1991). The HWL represents the landward extent of the most recent high tide and is characterised by a change in sand colour due to repeated, periodic inundation by high tides. The HWL is portrayed on aerial photographs by the most landward change in colour or grey tone (Boak & Turner 2005).

Importance and Application

The location of the shoreline and its changing position over time is of fundamental importance to coastal scientists, engineers and managers (Boak & Turner 2005; Pajack & Leatherman 2002). Present day shoreline monitoring campaigns provide information about historic shoreline location and movement, and about predictions of future change (Appeaning Addo et al. 2008). More specifically the position of the shoreline in the past, at present and where it is predicted to be in the future is useful for in the design of coastal protection, to calibrate and verify numerical modells to assess sea level rise, map hazard zones and formulate policies to regulate coastal development. Accurate and consistent delineation of the shoreline is integral to all of these tasks. The location of the shoreline also provides information regarding shoreline reorientation adjacent to structures, beach width, volume and rates of historical change (Boak & Turner 2005; Pajack & Leatherman 2002).

Data Sources

A variety of data sources are available for examining shoreline position however, the availability of historical data is limited at many coastal sites and so the choice of data source is largely limited to what is available for the site at a given time (Boak & Turner 2005). Shoreline mapping techniques applied to data sources have moved towards automation in association with technological advances and the need to reduce uncertainty. Although these changes have resulted in improvement in coastal data processing and storage capabilities, the frequent change in technology has prevented the emergence of one standard method of shoreline mapping. This has occurred because each data source and associated method have their own unique capabilities and shortcomings (Moore 2000). A number of the data sources used for shoreline mapping and their associated advantages and disadvantages.

Historical Maps

In the event that a study requires the shoreline position to be mapped before the development of aerial photographs, or if the location has poor photograph coverage it is necessary to employ historical maps in order to detail shoreline position (Moore 2000). The main advantage and reason for using historical maps is that they are able to provide a historic record that is not available from other data sources. Many potential errors however are associated with historical coastal maps and charts. Such errors may be associated with scale, datum changes, distortions from uneven shrinkage, stretching, creases, tears and folds, different surveying standards, different publication standards, and projection errors (Boak & Turner 2005). The severity of these errors depends on the accuracy standards met by each map and the physical changes that have occurred since the publication of the map (Anders & Byrnes 1991). The oldest reliable source of shoreline data in the United States dates back to the early to mid 19th century and is the U.S Coast and Geodetic Survey/National Ocean Service T-sheets (Morton 1991). In the United Kingdom, many maps and charts were deemed to be inaccurate until around 1750. The founding of the Ordnance Survey in 1791 has since improved the accuracy of the mapping.

Aerial Photographs

Aerial photographs have been used since the 1920's to provide topographical information about an area. They are therefore a good database for compilation of historical shoreline change maps. Aerial photographs are the most commonly used data source in shoreline mapping because many coastal areas have extensive aerial photo coverage therefore providing a valuable record of shoreline position (Moore 2000). In general, aerial photographs provide good spatial coverage of the coast however temporal coverage is very much site specific depending on the flight path of the aeroplane. A second disadvantage associated with aerial photography is that the interpretation of the shoreline position is subjective given the dynamic nature of the coastal environment. This combined with various distortions inherent in aerial photographs can lead to significant error levels (Moore 2000).

Object Space Displacements

Conditions outside of the camera can cause objects in an image to be displaced from their true ground position. Such conditions may include ground relief, camera tilt and atmospheric refraction.

Relief displacement is prominent when photographing a variety of elevations. This situation causes objects above ground level to be displaced outward from the centre of the photograph and objects below ground level to be displaced toward the centre of the image (Figure 2). The severity of the displacement is affected negatively with decreases in flight altitude and as radial distance from the centre of the photograph increases. This distortion can be minimised by photographing numerous swaths and creating a mosaic of the images. This technique will create a focus for the centre of each photograph where distortion is minimised. It is important to note that this error is not common in shoreline mapping is the relief is fairly constant. It is however important to consider when mapping cliffs (Moore 2000).

Ideally aerial photographs are taken so the optical axis of the camera is perfectly perpendicular to the ground surface thereby creating a vertical photograph. Unfortunately this is not often the case and virtually all aerial photographs experience tilt whereby up to 3° is not uncommon (Camfield et al. 1996). In this situation the scale of the image will be larger on the upward side of the tilt axis and smaller on the downward side. Moore, (2000) notes that many coastal researchers have not realised the severity of this error and therefore do not consider it in their methods.

Radial Lens Distortion

Lens distortion varies as a function of radial distance from the isocentre of the photograph meaning that the centre of the image is relatively distortion free, but as the angle of view increases the distortion becomes more prominent. This is a significant source of error in earlier aerial photography but as technology has increased and camera lens have become more refined it has become less of an issue with later photographs. Such a distortion is impossible to correct for without knowing the make and model of the lens used to capture the image. However if overlapping images have been acquired one can digitize the centre portions of the aerial photographs (Crowell et al. 1991).

Delineation of the Shoreline

The dynamic nature of the coast has meant that accurate mapping of an instantaneous shoreline position has been associated with significant uncertainty. This uncertainty arises because at any given time the position of the shoreline is influenced by the short-term effect of the tide and a wide variety of long term effects such as relative sea-level rise and along shore littoral sediment movement. Not only does this affect the accuracy of computed historic shoreline position but also any predicted future positions (Appeaning Addo et al. 2008). As mentioned earlier the HWL is most commonly used as a shoreline indicator. This can usually be seen as a significant tonal change on aerial photographs. There are however many errors associated with using the wet/dry line as a proxy for the HWL and shoreline. The errors of largest concern are the short term migration of the wet/dry line, interpretation of the wet/dry line on a photograph and measurement of the interpreted line position (Leatherman 2003; Moore 2000). Systematic errors such as the migration of the wet/dry line may arise from tidal and seasonal changes.

Storm-induced erosion is another factor which may cause the wet/dry line to migrate landward. Field investigations have shown that these changes can be minimised by using only summertime data (Moore 2000; Leatherman 2003). Furthermore, the error bar can be significantly reduced by using the longest record of reliable data to calculate erosion rates (Leatherman 2003). Finally it is important to note that errors may arise due to the difficulty of measuring a single line on a photograph. For example where the pen line is 0.13 mm thick this translates to an error of ±2.6 m on a 1:20000 scale photograph.

Beach Profiling Surveys

Beach profiling surveys are typically repeated at regular intervals along the coast in order to measure short-term (daily to annual) variations in shoreline position and beach volume. (Smith & Zarillo 1990). Beach profiling is a very accurate source of information however measurements are generally subject to the limitations of conventional surveying techniques. Shoreline data derived from beach profiling is often spatially and temporally limited due to the high cost associated with such a labour intensive activity. Shorelines are generally derived by interpolating between a series of discrete beach profiles. It is important to note however that the distance between the profiles is usually quite large and so the accuracy of the interpolating becomes compromised. In contrast to aerial photographs, survey data is limited to smaller lengths of shoreline generally less than ten kilometres (Boak & Turner 2005). Beach profiling data is commonly available in from regional councils in New Zealand such as those compiled by the Hawkes Bay Regional Council.

Remote Sensing

Technological advancement over the last decade has led to the development of a range of airborne, satellite and land based remote sensing techniques (Smith & Zarillo 1990). Some of the remotely sensed data sources are listed below:

- Multispectral and hyperspectral imaging
- Microwave sensors
- Global positioning system (GPS)
- Airborne light detection and ranging technology (LIDAR).

Remote sensing techniques are attractive as they are cost effective, reduce manual error and remove the subjective approach of conventional field techniques (Maiti et al. 2009). Remote sensing is a relatively new concept and so extensive historical observations are unavailable. Given this idea, it is important that coastal morphology observations are quantified by coupling remotely sensed data with other sources of information detailing historic shoreline position from archived sources (Appeaning Addo et al. 2008).

Video Analysis

Video analysis provides quantitative, cost-effective, continuous and long-term monitoring beaches (Turner et al. 2004). The advancement of coastal video systems over the past 15 years has resulted in the extraction of large amounts of geophysical data from images. Such data includes that about coastal morphology, surface currents and wave parametres. The main advantage of video analysis lies in the ability to reliably quantify these parametres with high resolution and coverage in both space and time. This in particular highlights their potential importance as an effective coastal monitoring system and an aid to coastal zone management (Van Koningsveld et al. 2007). Interesting case studies have been carried out using video analysis. Turner et al. (2004) used a video-based ARGUS coastal imaging system to monitor and quantify the regional-scale coastal response to sand nourishment and construction of the world-first Gold Coast artificial (surfing) reef in Australia. In addition, Smit et al. (2007) demonstrated the added value of high resolution video observations for making short-term predictions of near shore hydrodynamic and morphological processes, at temporal scales of metres to kilometres and days to seasons.

Chapter 2

Architectural Engineering

Architectural engineering, also known as building engineering, is the application of engineering principles and technology to building design and construction.

Figure: *César Pelli's Ratner Athletic Center uses cables, counterweights and masts as load-bearing devices.*

Definitions of an architectural engineer may refer to:

- An engineer in the structural, mechanical, electrical, construction or other engineering fields of building design and construction.
- A licensed engineering professional in parts of the United States.

- In informal contexts, and formally in some places, a professional synonymous with or similar to an architect. In some languages, "architect" is literally translated as "architectural engineer."

Engineering for Building

Structural engineering involves the analysis and design of physical objects such as buildings, bridges, equipment supports, towers and walls. Those concentrating on buildings are responsible for the structural performance of a large part of the built environment and are, sometimes, informally referred to as "building engineers". Structural engineers require expertise in strength of materials and in the seismic design of structures covered by earthquake engineering. Architectural Engineers sometimes practice structural as one aspect of their designs; the structural discipline when practised as a speciality works closely with architects and other engineering specialists.

Mechanical, Electrical, and Plumbing (MEP)

Figure: *MEP room in a building*

Mechanical engineering and Electrical engineering are specialists, commonly referred to as "MEP" (mechanical, electrical, and plumbing) when engaged in the building design fields. Also known as "building services engineering" in the United Kingdom, Canada, and Australia. Mechanical engineers design and oversee the heating ventilation and air conditioning (HVAC), plumbing, and rain gutter systems. Plumbing designers often include design specifications for simple active fire protection systems, but for more complicated projects, fire protection engineers are often separately retained. Electrical engineers are responsible for the building's power distribution, telecommunication, fire alarm, signalisation, lightning protection and control systems, as well as lighting systems.

The Architectural Engineer (PE) in the United States

In many jurisdictions of the United States, the architectural engineer is a licensed engineering professional, usually a graduate of an architectural engineering university program preparing students to perform whole-building design in competition with architect-engineer teams; or for practice in one of structural, mechanical or electrical fields of building design, but with an appreciation of integrated architectural requirements. Formal architectural engineering education, following the engineering model of earlier disciplines, developed in the late 19th century, and became widespread in the United States by the mid 20th century. With the establishment of a specific "architectural engineering" NCEES Professional Engineering registration examination in the 1990's, and first offering in April 2003, architectural engineering became recognised as a distinct engineering discipline in the United States. Architectural engineers are not entitled to practice architecture unless they are also licensed as architects.

The Architect as Architectural Engineer

In some countries architecture, as a profession providing architectural services, is sometimes referred to as "architectural engineering". In others, such as in Japan, the terms "architecture" and "building engineering" are used synonymously. The practice of architecture includes the planning, designing and overseeing the building's construction.

In some languages, such as Korean and Arabic, "architect" is literally translated as "architectural engineer". In some countries, an "architectural engineer" (such as the *ingegnere edile* in Italy) is entitled to practice architecture and is often referred to as an architect. These individuals are often also structural engineers. In other countries, such as Germany, Austria, Pakistan and most of the Arabic countries, architecture graduates receive an engineering degree (*Dipl.-Ing. - Diplom-Ingenieur*). In Brazil, architects and engineers currently share the same accreditation process (CREA - Regional Council of Engineers and Architects). Besides traditional architecture design training, Brazilian architecture courses also offer complementary training in engineering disciplines such as structural, electrical, hydraulic and mechanical engineering. After graduation, architects can be fully responsible for most engineering design and construction, except highly specialised tasks such as road design and high voltage electrical.

Education

The architectural, structural, mechanical and electrical engineering branches each have well established educational requirements that are usually fulfilled by completion of a university program.

Architectural Engineering as a Single Integrated Field of Study

What differentiates architectural engineering as a separate and single, integrated field of study, compared to other engineering disciplines, is its multi-disciplined engineering approach. Through training in and appreciation of architecture, the field seeks integration of building systems within its overall building design. Architectural engineering includes the design of building systems including Heating, ventilation and air conditioning (HVAC), plumbing, fire protection, electrical, lighting, architectural acoustics, and structural systems. In some university programs, students are required to concentrate on one of the systems; in others, they can receive a generalist architectural or building engineering degree.

Facade Engineering

Building facades make a major contribution to the overall aesthetic and technical performance of a building. Facade engineers operate within technical divisions of facade manufacturing companies while some engineers provide facade consultancy for architects, building owners, cladding manufacturers and construction managers. Projects can include new buildings and recladding of existing buildings.

Facade engineers work in conjunction with other members of the design team including fire, acoustic and lighting engineers. Their goal is to balance the overall objectives of appearance, quality, performance, cost, durability and schedule. Facade engineering requires a blend of skills ranging from structural engineering through to building physics, architecture, manufacturing, materials science, dynamics, programming, procurement, project management and construction techniques.

Fire Protection Engineering

Fire Protection Engineering (also known as fire engineering or fire safety engineering) is the application of science and engineering principles to protect people and their environments from the destructive effects of fire and smoke.

The discipline of fire protection engineering includes, but is not exclusive to:

- Active fire protection - fire suppression systems, and fire alarm.
- Passive fire protection - fire and smoke barriers, space separation
- Smoke control and management
- Escape facilities- Emergency exits, Fire lifts etc.
- Building design, layout, and space planning
- Fire prevention programs
- Fire dynamics and fire modelling
- Human behaviour during fire events
- Risk analysis, including economic factors
- Wildfire Management.

Fire protection engineers identify risks and design safeguards that aid in preventing, controlling, and mitigating the effects of fires. Fire protection engineers assist architects, building owners and developers in evaluating buildings' life safety and property protection goals. FPEs are also employed as fire investigators, including such very large-scale cases as the analysis of the collapse of the World Trade Centres. NASA uses fire protection engineers in its space program to help improve safety. Fire protection engineers are also employed to provide 3rd party review for performance based fire engineering solutions submitted in support of local building regulation applications.

History

Fire protection engineering (FPE) roots dating as far back as Ancient Rome, when the Emperor Nero ordered the city to be rebuilt utilising passive fire protection methods, such as space separation and noncombustible building materials, after a catastrophic fire.The discipline of fire protection engineering emerged in the early 20th century as a distinct discipline, separate from civil, mechanical and chemical engineering, in response to new fire problems posed by the Industrial Revolution. Fire protection engineers of this era concerned themselves with devising methods to protect large factories, particularly spinning mills and other manufacturing properties. Another motivation to organize the discipline, define practices and conduct research to support innovations was in response to the catastrophic conflagrations and mass urban fires that swept many major cities during the latter half of the 19th century. The insurance industry also helped promote advancements in the fire protection engineering profession and the development of fire protection systems and equipment.

In 1903 the first degree program in fire protection engineering was initiated as the Armour Institute of Technology (later becoming part of the Illinois Institute of Technology).

As the 20th century emerged, several catastrophic firesresulted in changes to buildings codes to better protect people and property from fire. It was only in the latter half of the 20th Century that fire protection engineering emerged as a unique engineering profession. The primary reason for this emergence was the development of the "body of knowledge," specific to the profession that occurred after 1950. Other factors contributing to the growth of the profession include the start of the Institution of Fire Engineers in 1918 in the UK, and the Society of Fire Protection Engineers in 1950 in the USA, the emergence of independent fire protection consulting engineers, and the promulgation of engineering standards for fire protection.

Education

Fire protection engineers, like their counterparts in other engineering and scientific disciplines, undertake a formal course of education and continuing professional development to acquire and maintain their competence. This education typically includes foundation studies in mathematics, physics, chemistry, and technical writing. Professional engineering studies focus students on acquiring proficiency in material science, statics, dynamics, thermodynamics, fluid dynamics, heat transfer, engineering economics, ethics, Systems in engineering, reliability, and environmental psychology. Studies in combustion, probabilistic risk assessment or risk management, the design of fire suppression systems, the application and interpretation of model building codes, and the measurement and simulation of fire phenomena complete most curricula.

New Zealand was one of the first countries in the world to introduce Performance based assessment methods into their building codes in regard to fire safety. This occurred with the introduction of their 1991 Building Act. Professor Andy Buchanan, of the University of Canterbury, established the first post graduate and only course available in New Zealand, at the time, in fire safety engineering in 1995. Applicants to the course require a minimum qualification of a bachelors degree in engineering or bachelors degree in a limited list of science course. Notable alumni from the university of Canterbury include but is not limited to Sir Ernest Rutherford, Robert (Bob) Park, Roy Kerr, Michael P. Collins, John Britten and many others. A masters degree in fire engineering from the University of Canterbury is recognised under the Washington Accord.

In the United States, the University of Maryland (UMD) offers the only ABET-accredited B.S. degree program in Fire Protection Engineering, as well as graduate degrees and a distance M.Eng. program. Worcester Polytechnic Institute (WPI) offers a M.S. and a Ph.D. in Fire Protection Engineering. Oklahoma State University offers a ABET-accredited B.S. in Fire Protection and Safety Engineering Technology (established in 1937), the University of Houston–Downtown (UHD) is ABET-accredited for a B.S. in Fire Protection Engineering Technology, and the University of Cincinnati offers an associate degree in Fire Science and a bachelor's degree in Fire and Safety Engineering Technology as distance learning options, the only university in the U.S. and Canada to hold this distinction.. Other institutions, such as the University of New Haven, University of Kansas, Illinois Institute of Technology, University of California, Berkeley, Eastern Kentucky University, and the University of Texas at Austin have offered courses in Fire Protection Engineering or technology.

The practice of final fire sprinkler systems design and hydraulic calculations is commonly performed by design technicians who are often educated in-house at contracting firms throughout North America, with the objective of preparing designers for certification by testing by associations such as NICET (National Institute for Certification in Engineering Technologies). NICET certification is commonly used as a proof of competency for securing a license to design and install fire protection systems.

In Europe, the University of Edinburgh offers a degree in Fire Engineering and had its first fire research group in the 1970's. These activities are now conducted at the new BRE Centre for Fire Safety Engineering. Other European Universities active in the fire engineering are Luleå tekniska universitet, Lund University, Stord/Haugesund University College, University of Central Lancashire, University of Manchester, University of Ulster, University of Sheffield, University of Leeds, University of Greenwich, London South Bank University, University of Wales, Newport, Letterkenny Institute of Technology in Ireland and Otto-von-Guericke-Universität Magdeburg and Bergische Universität Wuppertal in Germany.

In Australia, Victoria University in Melbourne offers postgraduate courses in Building Fire Safety and Risk Engineering as does the University of Western Sydney. The Centre for Environmental Safety and Risk Engineering (CESARE) is a research unit under Victoria University and has facilities for research and testing of fire behaviour.

Professional Registration

Suitably qualified and experienced fire protection engineers may qualify for registration as a professional engineer. The recognition of fire protection engineering as a separate discipline varies from state to state in the United States. Few countries outside the United States regulate the professional practice of fire protection engineering as a discipline, although they may restrict the use of the title engineer in association with its practice. The titles *fire engineer* and *fire safety engineer* tend to be preferred outside the United States, especially in the United Kingdom and Commonwealth countries influenced by the British fire service. The Institution of Fire Engineers is one international organisation that qualifies many aspects of the training and qualifications of fire engineers.

Roof

A roof is the covering on the uppermost part of a building. A roof protects the building and its contents from the effects of weather. Structures that require roofs range from a letter box to a cathedral or stadium, dwellings being the most numerous. In most countries a roof protects primarily against rain. Depending upon the nature of the building, the roof may also protect against heat, sunlight, cold, snow and wind. Other types of structure, for example, a garden conservatory, might use roofing that protects against cold, wind and rain but admits light. A verandah may be roofed with material that protects against sunlight but admits the other elements. The characteristics of a roof are dependent upon the purpose of the building that it covers, the available roofing materials and the local traditions of construction and wider concepts of architectural design and practice and may also be governed by local or national legislation.

Design Elements

The elements in the design of a roof are:

- the material
- the construction
- the durability.

The material of a roof may range from banana leaves, wheaten straw or seagrass to lamininated glass, aluminium sheeting and precast concrete. In many parts of the world ceramic tiles have been the predominant roofing material for centuries.

Figure: *The roofs of Olomouc, Czech Republic*

Figure: *The roofs (with domes) of the Great Mosque of Kairouan also called the Mosque of Uqba, located in Kairouan, Tunisia*

The construction of a roof is determined by its method of support and how the underneath space is bridged and whether or not the roof is *pitched*. The *pitch* is the angle at which the roof rises from its lowest to highest point. Most domestic architecture, except in very dry regions, has roofs that are sloped, or *pitched*. The pitch is partly dependent upon stylistic factors, but has more to do with practicalities. Some types of roofing, for example thatch, require a steep pitch in order to be waterproof and durable. Other types of roofing, for example pantiles, are unstable on a steeply pitched roof but provide excellent weather protection at a relatively low angle. In regions where there is little rain, an almost flat roof with a slight runoff provides adequate protection against an occasional downpour.

The durability of a roof is a matter of concern because the roof is often the least accessible part of a building for purposes of repair and renewal, while its damage or destruction can have serious effects.

Form of a Roof

The shape of roofs differs greatly from region to region. The main factors which influence the shape of roofs are the climate and the materials available for roof structure and the outer covering.

The basic shapes of roofs are flat, skillion, gabled, hipped, arched and domed. There are many variations on these types. Roofs constructed of flat sections that are sloped are referred to as pitched roofs (generally if the angle exceeds 10 degrees). Pitched roofs, including gabled, hipped and skillion roofs, make up the greatest number of domestic roofs. Some roofs follow organic shapes, either by architectural design or because a flexible material such as that is used in the construction.

Parts of a Roof

There are two parts to a roof, its supporting structure and its outer skin, or uppermost weatherproof layer. In a minority of buildings, the outer layer is also a self-supporting structure. The roof structure is generally supported upon walls, although some building styles, for example, geodesic and A-frame, blur the distinction between wall and roof.

Support

The supporting structure of a roof usually comprises beams that are long and of strong, fairly rigid material such as timber, and since the mid 19th century, cast iron or steel. In countries that use bamboo extensively, the flexibility of the material causes a distinctive curving line to the roof, characteristic of Oriental architecture. Timber lends itself to a great variety of roof shapes. The timber structure can fulfil an aesthetic as well as practical function, when left exposed to view.

Stone lintels have been used to support roofs since prehistoric times, but cannot bridge large distances. The stone arch came into extensive use in the ancient Roman period and in variant forms could be used to span spaces up to 140 feet (43 m) across. The stone arch or vault, with or without ribs, dominated the roof structures of major architectural works for about 2,000 years, only giving way to iron beams with the Industrial Revolution and the designing of such buildings as Paxton's Crystal Palace, completed 1851.

With continual improvements in steel girders, these became the major structural support for large roofs, and eventually for ordinary houses as well. Another form of girder is the reinforced concrete beam, in which metal rods are encased in concrete, giving it greater strength under tension.

Outer Layer

This part of the roof shows great variation dependent upon availability of material. In simple vernacular architecture, roofing material is often vegetation, such as thatches, the most durable being sea grass with a life of perhaps 40 years. In many Asian countries bamboo is used both for the supporting structure and the outer layer where split bamboo stems are laid turned alternately and overlapped. In areas with an abundance of timber, wooden shingles are used, while in some countries the bark of certain trees can be peeled off in thick, heavy sheets and used for roofing.

The 20th century saw the manufacture of composition shingles which can last from a thin 20 year shingle to the thickest which are limited lifetime shingles, the cost depending on the thickness and durability of the shingle. When a layer of shingles wears out, they are usually stripped, along with the underlay and roofing nails, allowing a new layer to be installed. An alternative method is to install another layer directly over the worn layer. While this method is faster, it does not allow the roof sheathing to be inspected and water damage, often associated with worn shingles, to be repaired. Having multiple layers of old shingles under a new layer causes roofing nails to be located further from the sheathing, weakening their hold. The greatest concern with this method is that the weight of the extra material could exceed the dead load capacity of the roof structure and cause collapse.

Slate is an ideal, and durable material, while in the Swiss Alps roofs are made from huge slabs of stone, several inches thick. The slate roof is often considered the best type of roofing. A slate roof may last 75 to 150 years, and even longer. However, slate roofs are often expensive to install – in the USA, for example, a slate roof may have the same cost as the rest of the house. Often, the first part of a slate roof to fail is the fixing nails; they corrode, allowing the slates to slip. In the UK, this condition is known as "nail sickness". Because of this problem, fixing nails made of stainless steel or copper are recommended, and even these must be protected from the weather.

Asbestos, usually in bonded corrugated panels, has been used widely in the 20th century as an inexpensive, nonflammable roofing material with excellent insulating properties. Health and legal issues involved in the mining and handling of asbestos products means that it is no longer used as a new roofing material. However, many asbestos roofs continue to exist, particularly in South America and Asia. Roofs made of cut turf (modern ones known as Green roofs, traditional ones as sod roofs) have good insulating properties and are increasingly encouraged as a way of "greening" the Earth. Adobe roofs are roofs of clay, mixed with binding material such as straw or animal hair, and plastered on lathes to form a flat or gently sloped roof, usually in areas of low rainfall.

In areas where clay is plentiful, roofs of baked tiles have been the major form of roof. The casting and firing of roof tiles is an industry that is often associated with brickworks. While the shape and colour of tiles was once regionally distinctive, now tiles of many shapes and colours are produced commercially, to suit the taste and pocketbook of the purchaser.

Sheet metal in the form of copper and lead has also been used for many hundreds of years. Both are expensive but durable, the vast copper roof of Chartres Cathedral, oxidised to a pale green colour, having been in place for hundreds of years. Lead, which is sometimes used for church roofs, was most commonly used as flashing in valleys and around chimneys on domestic roofs, particularly those of slate. Copper was used for the same purpose.

In the 19th century, iron, electroplated with zinc to improve its resistance to rust, became a lightweight, easily-transported, waterproofing material. Its low cost and easy application made it the most accessible commercial roofing, world wide. Since then, many types of metal roofing have been developed. Steel shingle or standing-seam roofs last about 50 years or more depending on both the method of installation and the moisture barrier (underlayment) used and are between the cost of shingle roofs and slate roofs. In the 20th century a large number of roofing materials were developed, including roofs based on bitumen (already used in previous centuries), on rubber and on a range of synthetics such as thermoplastic and on fibreglass.

Tower

A tower is a tall structure, usually taller than it is wide, often by a significant margin. Towers are distinguished from masts by their lack of guy-wires.

The Eiffel Tower in Paris, France.

The Two Towers in Bologna, Italy.

Towers are generally built to take advantage of their height, and can stand alone on the ground, or as part of a larger structure or device such as a fortified building or as an integral part of a bridge, the term also denoting a raised structure on a ship or other vehicle.

History

Towers have been used by mankind since prehistoric times. The oldest known may be the circular stone tower in walls of Neolithic Jericho (8000 BC). Some of the earliest towers were ziggurats, which existed in Sumerian architecture since the 4th millennium BC. The most famous ziggurats include the Sumerian Ziggurat of Ur, built the 3rd millennium BC, and the Etemenanki, one of the most famous examples of Babylonian architecture. The latter was built in Babylon during the 2nd millennium BC and was considered the tallest tower of the ancient world.

Some of the earliest surviving examples are the broch structures in northern Scotland, which are conical towerhouses. These and other examples from Phoenician and Roman cultures emphasized the use of a tower in fortification and sentinel roles. For example, watchtower elements are found at Mogador from the first millennium BC, derived from Phoenician or Carthaginian origins.

The Romans utilised octagonal towers as elements of Diocletian's Palace in Croatia, which monument dates to approximately 300 AD, while the Servian Walls (4th century BC) and the Aurelian Walls (3rd century AD) featured square ones. The Chinese used towers as integrated elements of the Great Wall of China in 210 BC during the Qin Dynasty. Towers were also an important element of castles.

Another well known tower is the Leaning Tower of Pisa in Pisa, Italy built from 1173 until 1372. The Himalayan Towers are stone towers located chiefly in Tibet built approximately 14th to 15th century.

Mechanics

Up to a certain height, a tower can be made with the supporting structure with parallel sides. However, above a certain height, the compressive load of the material is exceeded and the tower will fail. This can be avoided if the tower's support structure tapers up the building.

A second limit is that of buckling- the structure requires sufficient stiffness to avoid breaking under the loads it faces, especially those due to winds. Many very tall towers have their support structures at the periphery of the building, which greatly increases the overall stiffness.

A third limit is dynamic; a tower is subject to varying winds, vortex shedding, seismic disturbances etc. These are often dealt with a combination of simple strength and stiffness, as well as in some cases tuned mass dampers to damp out movements. Varying or tapering the outer aspect of the tower with height avoids vibrations due to vortex shedding occurring along the entire building simultaneously.

Functions

Skyscrapers: A modern type of tower, the skyscraper, uses less ground space as a ratio of total building interior square footage. Skyscrapers are often not classified as towers, although most have the same design and structure of towers. In the United Kingdom, tall domestic buildings are referred to as tower blocks. In the United States, the World Trade Centre had the nickname the Twin Towers, a name shared with the Petronas Twin Towers in Kuala Lumpur. A tower's has very deep foundation's.

Strategic Advantages

The tower throughout history has provided its users with an advantage in surveying defensive positions and obtaining a better view of the surrounding areas, including battlefields. They were installed on defensive walls, or rolled near a target. Today, strategic-use towers are still used at prisons, military camps, and defensive perimetres.

Potential Energy

By using gravity to move objects or substances downward, a tower can be used to store items or liquids like a storage silo or a water tower, or aim an object into the earth such as a drilling tower. Ski-jump ramps use the same idea, and in the absence of a natural mountain slope or hill, can be human-made.

Communication Enhancement

In history, simple towers like lighthouses, bell towers, clock towers, signal towers and minarets were used to communicate information over greater distances. In more recent years, radio masts and cell phone towers facilitate communication by expanding the range of the transmitter. The CN Tower in Toronto, Canada was built as a communications tower, with the capability to act as both a transmitter and repeater. Its design also incorporated features to make it a tourist attraction, including the world's highest observation deck at 147 stories.

Transportation Support

Towers can also be used to support bridges, and can reach heights that rival some of the tallest buildings above-water. Their use is most prevalent in suspension bridges and cable-stayed bridges. The use of the pylon, a simple tower structure, has also helped to build railroad bridges, mass-transit systems, and harbours. Control towers are used to give visibility to help direct aviation traffic.

Other Towers

- To access tall or high objects: launch tower, service tower, service structure, scaffold, tower wagon
- To access atmospheric conditions aloft: wind turbine, meteorological measurement tower, tower telescope, solar power station

- To lift high tension cables for electrical power distribution transmission tower
- To take advantage of the temperature gradient inherent in a height differential: cooling tower, chimney
- To protect from exposure: BREN Tower
- For industrial production: shot tower
- For surveying: Survey tower
- To drop objects: Drop tube (drop tower), bomb tower, diving platform
- To test height-intensive applications: elevator test tower
- To improve structural integrity: thyristor tower
- To mimic towers or provide height for training purposes: fire tower, parachute tower
- As art: Shukhov Tower
- For recreation: rock climbing tower
- As a symbol: Tower of Babel, The Tower (Tarot card), church tower

The term "tower" is also sometimes used to refer to firefighting equipment with an extremely tall ladder designed for use in firefighting/ rescue operations involving high-rise buildings.

Wind Engineering

Wind engineering analyses effects of wind in the natural and the built environment and studies the possible damage, inconvenience or benefits which may result from wind. In the field of structural engineering it includes strong winds, which may cause discomfort, as well as extreme winds, such as in a tornado, hurricane or heavy storm, which may cause widespread destruction. In the fields of wind energy and air pollution it also includes low and moderate winds as these are relevant to electricity production resp. dispersion of contaminants.

Wind engineering draws upon meteorology, fluid dynamics, mechanics, Geographic Information Systems and a number of specialist engineering disciplines including aerodynamics, and structural dynamics. The tools used include atmospheric modells, atmospheric boundary layer wind tunnels and computational fluid dynamics modells.

Wind engineering involves, among other topics:

- Wind impact on structures (buildings, bridges, towers).
- Wind comfort near buildings.

- Effects of wind on the ventilation system in a building.
- Wind climate for wind energy.
- Air pollution near buildings.

Wind engineering may be considered by structural engineers to be closely related to earthquake engineering and explosion protection.

History

Wind Engineering as a separate discipline can be traced to the UK in the 1960's, when informal meetings were held at the [National Physical Laboratory (United Kingdom) | [National Physical Laboratory]], the Building Research Establishment and elsewhere.

Wind Loads on Buildings

The design of buildings must account for wind loads, and these are affected by wind shear. For engineering purposes, a power law wind speed profile may be defined as follows:

$$v_z = v_g \cdot \left(\frac{z}{z_g}\right)^{\frac{1}{\alpha}}, 0 < z < z_g$$

where:

v_z = speed of the wind at height z

v_g = gradient wind at gradient height z_g

α = exponential coefficient.

Typically, buildings are designed to resist a strong wind with a very long return period, such as 50 years or more. The design wind speed is determined from historical records using Extreme value theory to predict future extreme wind speeds.

Wind Turbines

Wind turbines are affected by wind shear. Vertical wind-speed profiles result in different wind speeds at the blades nearest to the ground level compared to those at the top of blade travel, and this in turn affects the turbine operation. The wind gradient can create a large bending moment in the shaft of a two bladed turbine when the blades are vertical. The reduced wind gradient over water means shorter and less expensive wind turbine towers can be used in shallow seas.

For wind turbine engineering, wind speed variation with height is often approximated using a power law:

$$v_w(h) = v_{ref} \cdot \left(\frac{h}{h_{ref}} \right)^a$$

where:

$v_w(h)$= velocity of the wind at height h [m/s]

v_{ref}= velocity of the wind at some reference height h_{ref} [m]

a = Hellman exponent (aka power law exponent or shear exponent) (~= 1/7 in neutral flow, but can be >1)

Significance

The knowledge of wind engineering is used to analyse and design all high rise buildings, cable suspension bridges and cable-stayed bridges, electricity transmission towers and telecommunication towers and all other types of towers and chimneys. The wind load is the dominant load in the analysis of many tall buildings. So wind engineering is essential for the analysis and design of tall buildings. Again, wind load is a dominant load in the analysis and design of all long-span cable bridges.

Surveying

Surveying or land surveying is the technique, profession, and science of accurately determining the terrestrial or three-dimensional position of points and the distances and angles between them. These points are usually on the surface of the Earth, and they are often used to establish land maps and boundaries for ownership or governmental purposes. To accomplish their objective, surveyors use elements of mathematics (geometry and trigonometry), physics, engineering and law.

An alternative definition, per the American Congress on Surveying and Mapping (ACSM), is the science and art of making all essential measurements to determine the relative position of points and/or physical and cultural details above, on, or beneath the surface of the Earth, and to depict them in a usable form, or to establish the position of points and/or details.

Furthermore, as alluded to above, a particular type of surveying known as "land surveying" (also per ACSM) is the detailed study or inspection, as by gathering information through observations, measurements in the field, questionnaires, or research of legal instruments, and data analysis in the support of planning, designing, and establishing of property boundaries. It involves the re-establishment of cadastral surveys and land boundaries based on documents of record and historical evidence, as well as certifying surveys (as required by statute or local ordinance) of subdivision plats/ maps, registered land surveys, judicial surveys, and space delineation. Land surveying can include associated services such as mapping and related data accumulation, construction layout surveys, precision measurements of length, angle, elevation, area, and volume, as well as horizontal and vertical control surveys, and the analysis and utilisation of land survey data. Surveying has been an essential element in the development of the human environment since the beginning of recorded history (about 5,000 years ago). It is required in the planning and execution of nearly every form of construction. Its most familiar modern uses are in the fields of transport, building and construction, communications, mapping, and the definition of legal boundaries for land ownership.

History of Surveying

Surveying techniques have existed throughout much of recorded history. In ancient Egypt, when the Nile River overflowed its banks and washed out farm boundaries, boundaries were re-established by a rope stretcher, or surveyor, through the application of simple geometry. The nearly perfect squareness and north-south orientation of the Great Pyramid of Giza, built c. 2700 BC, affirm the Egyptians' command of surveying.

A brief history of surveying:

- The Egyptian land register (3000 BC).
- A recent reassessment of Stonehenge (c. 2500 BC) suggests that the monument was set out by prehistoric surveyors using peg and rope geometry.
- The Groma surveying instrument originated in Mesopotamia (early 1st millennium BC).
- Under the Romans, land surveyors were established as a profession, and they established the basic measurements under which the Roman Empire was divided, such as a tax register of conquered lands (300 AD).
- The rise of the Caliphate led to extensive surveying throughout the Arab Empire. Arabic surveyors invented a variety of specialised instruments for surveying, including:
    - o Instruments for accurate levelling: A wooden board with a plumb line and two hooks, an equilateral triangle with a plumb line and two hooks, and a reed level.

  - o A rotating alhidade, used for accurate alignment.
  - o A surveying astrolabe, used for alignment, measuring angles, triangulation, finding the width of a river, and the distance between two points separated by an impassable obstruction.
- In England, The Domesday Book by William the Conqueror (1086)
  - o covered all England
  - o contained names of the land owners, area, land quality, and specific information of the area's content and inhabitants.
  - o did not include maps showing exact locations.

In the 18th century in Europe triangulation was used to build a hierarchy of networks to allow point positioning within a country. Highest in the hierarchy were triangulation networks. These were densified into networks of traverses (polygons), into which local mapping surveying measurements, usually with measuring tape, corner prism and the familiar red and white poles, are tied. For example, in the late 1970's, a team from the Ordnance Survey of Great Britain, originally under General William Roy began the Principal Triangulation of Britain using the specially built Ramsden theodolite. Large scale surveys are known as geodetic surveys.

- Continental Europe's Cadastre was created in 1808
  - o founded by Napoleon I (Bonaparte)
  - o contained numbers of the parcels of land (or just land), land usage, names etc., and value of the land
  - o 100 million parcels of land, triangle survey, measurable survey, map scale: 1:2500 and 1:1250
  - o spread fast around Europe, but faced problems especially in Mediterranean countries, Balkan, and Eastern Europe due to cadastre upkeep costs and troubles.

A cadastre loses its value if register and maps are not constantly updated. Because of the fundamental value of land and real estate to the local and global economy, land surveying was one of the first professions to require Professional Licensure. In many jurisdictions, the land surveyors license was the first Professional Licensure issued by the state, province, or federal government.

Surveying Techniques

Historically, distances were measured using a variety of means, such as with chains having links of a known length, for instance a Gunter's chain, or measuring tapes made of steel or invar. To measure horizontal distances, these chains or tapes were pulled taut according to temperature, to reduce sagging and slack. Additionally, attempts to hold the measuring instrument level would be made. In instances of measuring up a slope, the surveyor might have to "break" (break chain) the measurement- use an increment less than the total length of the chain.

Historically, horizontal angles were measured using a compass, which would provide a magnetic bearing, from which deflections could be measured. This type of instrument was later improved, with more carefully scribed discs providing better angular resolution, as well as through mounting telescopes with reticules for more-precise sighting atop the disc. Additionally, levels and calibrated circles allowing measurement of vertical angles were added, along with verniers for measurement to a fraction of a degree—such as with a turn-of-the-century transit.

The simplest method for measuring height is with an altimetre, basically a barometre, using air pressure as an indication of height. But surveying requires greater precision. A variety of means, such as precise levels (also known as differential levelling), have been developed to do this. With precise levelling, a series of measurements between two points are taken using an instrument and a measuring rod. Differentials in height between the measurements are added and subtracted in a series to derive the net difference in elevation between the two endpoints of the series. With the advent of the Global Positioning System (GPS), elevation can also be derived with sophisticated satellite receivers, but usually with somewhat less accuracy than with traditional precise levelling. However, the accuracies may be similar if the traditional levelling would have to be run over a long distance.

Triangulation is another method of horizontal location made almost obsolete by GPS. With the triangulation method, distances, elevations and directions between objects at great distance from one another can be determined. Since the early days of surveying, this was the primary method of determining accurate positions of objects for topographic maps of large areas. A surveyor first needs to know the horizontal distance between two of the objects. Then the height, distances and angular position of other objects can be derived, as long as they are visible from one of the original objects. High-accuracy transits or theodolites were used for this work, and angles between objects were measured repeatedly for increased accuracy.

Turning is a term used when referring to moving the level to take a elevation shot in a different location. When land surveying, there may be trees or other obstructions blocking the view from the level gun to the level rod. In order to "turn" the level gun, you mush first take a shot on the rod from your current location and record the elevation. Keeping the level rod in exactly the same location and elevation you may move the level gun to a different location where the level rod is still visible. Record the new elevation seen from the new location of the level rod and use the difference in elevations to find the new elevation of the level gun. Turning is not only used when there are obstructions in the way, but also when drastically changing elevations. You can turn up or down in elevation but the gun must always be at a higher elevation than the base of the rod. A level rod can usually be raised up to 25 feet high which enables the gun to be set much higher. However, if the gun is lower than the base of the rod, you will not be able to take a shot because the rod cannot be lowered beyond the ground elevation.

Surveying Equipment

As late as the 1990's, the basic tools used in planar surveying were a tape measure for determining shorter distances, a level to determine height or elevation differences, and a theodolite, set on a tripod, to measure angles (horizontal and vertical), combined with the process of triangulation. Starting from a position with known location and elevation, the distance and angles to the unknown point are measured.

A more modern instrument is a total station, which is a theodolite with an electronic distance measurement device (EDM). A total station can also be used for levelling when set to the horizontal plane. Since their introduction, total stations have made the technological shift from being optical-mechanical devices to being fully electronic.

Modern top-of-the-line total stations no longer require a reflector or prism (used to return the light pulses used for distancing) to return distance measurements, are fully robotic, and can even e-mail point data to the office computer and connect to satellite positioning systems, such as a Global Positioning System. Though real-time kinematic GPS systems have increased the speed of surveying, they are still horizontally accurate to only about 20 mm and vertically accurate to about 30–40 mm.

Total stations are still used widely, along with other types of surveying instruments. However, GPS systems do not work well in areas with dense tree cover or constructions. One-person robotic-guided total stations allow surveyors to gather precise measurements without extra workers to look through and turn the telescope or record data. A faster but expensive way to measure large areas (not details, and no obstacles) is with a helicopter, equipped with a laser scanner, combined with a GPS to determine the position and elevation of the helicopter. To increase precision, surveyors place beacons on the ground (about 20 km (12 mi) apart). This method reaches precisions between 5–40 cm (depending on flight height).

Types of Surveys and Applicability

- *ALTA/ACSM Survey*: a surveying standard jointly proposed by the American Land Title Association and the American Congress on Surveying and Mapping that incorporates elements of the boundary survey, mortgage survey, and topographic survey.
- *Archaeological survey*: used to accurately assess the relationship of archaeological sites in a landscape or to accurately record finds on an archaeological site.
- *As-built survey*: a survey carried out during or immediately after a construction project for record, completion evaluation and payment purposes. An as-built survey also known as a 'works as executed survey' documents the location of the recently constructed elements that are subject to completion evaluation. As built surveys are typically presented in red or redline and overlayed over existing design plans for direct comparison with design information.
- *Bathymetric survey:* a survey carried out to map the topography and features of the bed of an ocean, lake, river or other body of water.
- *Boundary survey*: a survey that establishes boundaries of a parcel using its legal description, which typically involves the setting or restoration of monuments or markers at the corners or along the lines of the parcel, often in the form of iron rods, pipes, or concrete monuments in the ground, or nails set in concrete or asphalt.
- *Deformation survey:* a survey to determine if a structure or object is changing shape or moving. The three-dimensional

positions of specific points on an object are determined, a period of time is allowed to pass, these positions are then re-measured and calculated, and a comparison between the two sets of positions is made.

- *Engineering surveys*: those surveys associated with the engineering design (topographic, layout and as-built) often requiring geodetic computations beyond normal civil engineering practise.
- *Foundation survey*: a survey done to collect the positional data on a foundation that has been poured and is cured. This is done to ensure that the foundation was constructed in the location, and at the elevation, authorised in the *plot plan*, *site plan*, or *subdivision plan*.
- *Geological survey*: generic term for a survey conducted for the purpose of recording the geologically significant features of the area under investigation..
- *Hydrographic survey*: a survey conducted with the purpose of mapping the coastline and seabed for navigation, engineering, or resource management purposes.
- *Measured survey* : a building survey to produce plans of the building. Such a survey may be conducted before renovation works, for commercial purpose, or at end of the construction process "as built survey"
- *Mortgage survey or physical survey*: a simple survey that delineates land boundaries and building locations. In many places a mortgage survey is required by lending institutions as a precondition for a mortgage loan.
- *Soil survey*, or soil mapping, is the process of determining the soil types or other properties of the soil cover over a landscape, and mapping them for others to understand and use.
- *Structural survey*: a detailed inspection to report upon the physical condition and structural stability of a building or other structure and to highlight any work needed to maintain it in good repair.
- *Tape survey*: this type of survey is the most basic and inexpensive type of land survey. Popular in the middle part of the 20th century, tape surveys while being accurate for distance lack substantially in their accuracy of measuring angle and bearing. Standards that are practised by professional land surveyors.

- *Topographic survey*: a survey that measures the elevation of points on a particular piece of land, and presents them as contour lines on a plot.

Surveying as a Career

The basic principles of surveying have changed little over the ages, but the tools used by surveyors have evolved tremendously. Engineering, especially civil engineering, depends heavily on surveyors.

Whenever there are roads, railways, reservoir, dams, retaining walls, bridges or residential areas to be built, surveyors are involved. They establish the boundaries of legal descriptions and the boundaries of various lines of political divisions. They also provide advice and data for *geographical information systems* (GIS), computer database that contain data on land features and boundaries.

Surveyors must have a thorough knowledge of algebra, basic calculus, geometry, and trigonometry. They must also know the laws that deal with surveys, property, and contracts.

In addition, they must be able to use delicate instruments with accuracy and precision. In the United States, surveyors and civil engineers use units of feet wherein a survey foot is broken down into 10th and 100th. Many deed descriptions requiring distance calls are often expressed using these units (125.25 ft). On the subject of accuracy, surveyors are often held to a standard of one one-hundredth of a foot; about 1/8 inch. Calculation and mapping tolerances are much smaller wherein achieving near-perfect closures are desired. Though tolerances such as this will vary from project to project, in the field and day to day usage beyond a 100th of a foot is often impractical.

In most of the United States, surveying is recognised as a distinct profession apart from engineering. Licensing requirements vary by state, but they generally have components of education, experience and examinations. In the past, experience gained through an apprenticeship, together with passing a series of state-administered examinations, was required to attain licensure. Now, most states insist upon basic qualification of a degree in surveying, plus experience and examination requirements.

The licensing process typically follows two phases. First, upon graduation, the candidate may be eligible to take the Fundamentals of Land Surveying exam, to be certified upon passing and meeting all other requirements as a surveyor in training (SIT). Upon being certified as an SIT, the candidate then needs to gain additional experience to become eligible for the second phase. That typically consists of the Principles and Practice of Land Surveying exam along with a state-specific examination.

Licensed surveyors usually denote themselves with the letters P.S. (professional surveyor), L.S. (land surveyor), P.L.S. (professional land surveyor), R.L.S. (registered land surveyor), R.P.L.S. (Registered Professional Land Surveyor), or P.S.M. (professional surveyor and mapper) following their names, depending upon the dictates of their particular jurisdiction of registration.

In Canada, land Surveyors are registered to work in their respective province. The designation for a land surveyor breaks down by province, but follows the rule whereby the first letter indicates the province, followed by L.S. There is also a designation as a C.L.S. or Canada lands surveyor, who has the authority to work on Canada Lands, which include Indian Reserves, National Parks, the three territories and offshore lands. In many Commonwealth countries, the term Chartered Land Surveyor is used for someone holding a professional license to conduct surveys. A licensed land surveyor is typically required to sign and seal all plans, the format of which is dictated by their state jurisdiction, which shows their name and registration number. In many states, when setting boundary corners land surveyors are also required to place survey monuments bearing their registration numbers, typically in the form of capped iron rods, concrete monuments, or nails with washers.

Building Surveying

Building surveying emerged in the 1970's as a profession in the United Kingdom by a group of technically minded General Practice Surveyors. Building surveying is a recognised profession in Britain, Ireland, Australia and Hong Kong. In Australia in particular, due to risk mitigation and limitation factors, the employment of surveyors at all levels of the construction industry is widespread. There are still many countries where it is not widely recognised as a profession.

Building Surveyors are trained to some extent in all aspects of property but with specific training in Building Pathology, as such they have a wide understanding of the end implications of decisions taken by more specific professions and trades during the realisation process, thus making them ideal Project and Property Managers on the client side. Services that building surveyors undertake are broad but can include:

- Construction design and building works
- Project management and monitoring
- Property Legislation advice
- Insurance assessment and claims assistance
- Defect investigation and maintenance advice
- Building surveys and measured surveys
- Handling planning applications
- Building inspection to ensure compliance with building regulations
- Pre-acquisition surveys
- Negotiating dilapidations claims.

Building surveyors also advise on many aspects of construction including:

- design
- cost
- maintenance
- sustainability
- repair
- refurbishment
- restoration and preservation of buildings and monuments.

Clients of a building surveyor can be the government agencies, businesses and individuals. Surveyors work closely with architects, planners, quantity surveyors, engineers, homeowners and tenants groups. A building surveyor may be called to act as an expert witness. It is usual for building surveyors to earn a college degree before undertaking structured training to become a member of a professional organisation. With the enlargement of the European community, the profession of the building surveyor is becoming more widely known in other European states, particularly France, where many English-speaking people buy second homes.

Lidar Surveying – Three-dimensional laser scanning provides high definition surveying for architectural, as-built, and engineering surveys. Recent technological advances make it the most cost-effective and time-sensitive solution for providing the highest level of detail available for interior and exterior building work.

Land Surveyor

The job of the land surveyor is to find and mark certain locations on land. A typical location of interest, for example, is the boundary of a person's property. That boundary is described in legal documents and the land surveyor follows that description and locates the boundary on the physical land and marks it, so the owner knows what land he can legally use. As an example, such a legal description may refer to a point as being 120.25 feet south of some existing marker. The land surveyor in that case would find the existing marker and use measuring instruments to find the point 120.25 feet south of that, and place a new marker at that location. These markers are called monuments.

Over time, development, vandalism, and acts of nature often wreak havoc on monumentation, so the land surveyor is often forced to consider other evidence such as fence locations, woodlines, monuments on neighboring property, recollections of people, and other evidence.

Reference monumentation refers to actual physical points on the ground that define location of boundary lines that divide neighboring parcels as well as their respective corners. Also called survey control, they are most often 1/2" or 5/8" iron rebar rods or pipes placed at 18" minimum depth. These rods and/or pipes usually have an affixed plastic cap over the top bearing the responsible surveyors' name and license number. In addition to rods and pipes, surveyors often use 4x4" concrete posts at corners of large parcels or anywhere that would require more stability (e.g. beach sand). They place them three feet deep. In places where there is asphalt or concrete, it is common to place nails or aluminum alloy caps to re-establish boundary corners. Marks are meant to be durable, stable, and as "permanent" as possible. The aim is to provide sufficient marks so some marks will remain for future re-establishment of boundaries. The material and marking used on monuments placed to mark boundary corners are often subject to state laws. Cadastral land surveyors are licensed by governments. In the United States, cadastral surveys are typically conducted by the federal government, specifically through the Cadastral Surveys branch of the Bureau of Land Management (BLM), formerly the General Land Office (GLO). They consult with USFS, Park Service, Corps of Engineers, BIA, Fish and Wildlife Service, Bureau of Reclamation, etc. In states that have been organised per the Public Land Survey System (PLSS), surveyors carry out BLM Cadastral Surveys in accordance with that system.

A common use of a survey is to determine a legal property boundary. The first stage in such a survey is to research relevant title records such as deeds, survey monumentation (marks on the ground), and any public or private records that provide relevant data.

In order to properly establish the position for survey markers, the surveyor must then take measurements. To do this, the surveyor usually places a total station over various points on the ground and records distances taken with the EDM. The surveyor analyses the data and makes comparisons with existing records to determine evidence that can be used to establish boundary positions. The surveyor calculates the bearing and distance of lines between the boundary corners and total station positions and uses them to set out and mark the corners in the field. He may check measurements by measuring directly between places using a flexible tape.

The Art of Surveying

Many properties have considerable problems with regards to improper bounding, miscalculations in past surveys, titles, easements, and wildlife crossings. Also many properties are created from multiple divisions of a larger piece over the course of years, and with every additional division the risk of miscalculation increases. The result can be abutting properties not coinciding with adjacent parcels, resulting in hiatuses (gaps) and overlaps. The art plays a role when a surveyor must solve a puzzle using pieces that do not exactly fit together. In these cases, the solution is based upon the surveyor's research and interpretation, along with established procedures for resolving discrepancies.

Chapter 3

Construction Surveying

Construction surveying (otherwise known as "layout" or "setting-out") is to stake out reference points and markers that will guide the construction of new structures such as roads or buildings. These markers are usually staked out according to a suitable coordinate system selected for the project.

History of Construction Surveying

- The nearly perfect squareness and north-south orientation of the Great Pyramid of Giza, built c. 2700 BC, affirm the Egyptians' command of surveying.
- A recent reassessment of Stonehenge (c. 2500 BC) suggests that the monument was set out by prehistoric surveyors using peg and rope geometry.
- In the sixth century BC geometric based techniques were used to construct the tunnel of Eupalinos on the island of Samos.

Elements of the Construction Survey

- Survey existing conditions of the future work site, including topography, existing buildings and infrastructure, and underground infrastructure whenever possible (for example, measuring invert elevations and diametres of sewers at manholes);
- Stake out reference points and markers that will guide the construction of new structures;
- Verify the location of structures during construction;
- Conduct an As-Built survey: a survey conducted at the end of the construction project to verify that the work authorised was completed to the specifications set on plans.

Coordinate Systems used in Construction

Land surveys and surveys of existing conditions are generally performed according to geodesic coordinates. However for the purposes of construction a more suitable coordinate system will often be used. During construction surveying, the surveyor will often have to convert from geodesic coordinates to the coordinate system used for that project.

Chainage or Station

In the case of roads or other linear infrastructure, a *chainage* will be established, often to correspond with the centre line of the road. During construction, structures would then be located in terms of *chainage*, *offset* and *elevation*. *Offset* is said to be "left" or "right" relative to someone standing on the *chainage line* who is looking in the direction of increasing *chainage*. Plans would often show *plan* views (viewed from above), *profile* views (a "transparent" section view collapsing all section views of the road parallel to the *chainage*) or *cross-section* views (a "true" section view perpendicular to the *chainage*). In a *plan* view, *chainage* generally increases from left to right, or from the bottom to the top of the plan. *Profiles* are shown with the chainage increasing from left to right, and *cross-sections* are shown as if the viewer is looking in the direction of increasing *chainage* (so that the "left" *offset* is to the *left* and the "right" *offset* is to the *right*).

"Chainage" may also be referred to as "Station".

Building Grids

In the case of buildings, an arbitrary system of grids is often established so as to correspond to the rows of columns and the major load-bearing walls of the building. The grids may be identified alphabetically in one direction, and numerically in the other direction (as in a road map). The grids are usually but not necessarily perpendicular, and are often but not necessarily evenly spaced. Floors and basement levels are also numbered. Structures, equipment or architectural details may be located in reference to the floor and the nearest intersection of the arbitrary axes.

Other Coordinate Systems

In other types of construction projects, arbitrary "north-south" and "east-west" reference lines may be established, that do not necessarily correspond to true coordinates.

Equipment and Techniques used in Construction Surveying

Surveying equipment, such as levels and theodolites, are used for accurate measurement of angular deviation, horizontal, vertical and slope distances. With computerisation, electronic distance measurement (EDM), total stations, GPS surveying and laser scanning have supplemented (and to a large extent supplanted) the traditional optical instruments. The builder's level measures neither horizontal nor vertical angles. It simply combines a spirit level and telescope to allow the user to visually establish a line of sight along a level plane. When used together with a graduated staff it can be used to transfer elevations from one location to another.

An alternative method to transfer elevation is to use water in a transparent hose as the level of the water in the hose at opposite ends will be at the same elevation.

Equipment and Techniques used in Mining and Tunneling

Total stations are the primary survey instrument used in mining surveying.

Underground Mining

A total station is used to record the absolute location of the tunnel walls (stopes), ceilings (backs), and floors as the drifts of an underground mine are driven. The recorded data is then downloaded into a CAD program, and compared to the designed layout of the tunnel.

The survey party installs control stations at regular intervals. These are small steel plugs installed in pairs in holes drilled into walls or the back. For wall stations, two plugs are installed in opposite walls, forming a line perpendicular to the drift. For back stations, two plugs are installed in the back, forming a line parallel to the drift.

A set of plugs can be used to locate the total station set up in a drift or tunnel by processing measurements to the plugs by intersection and resection.

Professional Status of Construction Surveyors

Building Surveying emerged in the 1970's as a profession in the United Kingdom by a group of technically minded General Practice Surveyors. Building Surveying is a recognised profession within Britain and Australia.

In Australia in particular, due to risk mitigation/limitation factors the employment of surveyors at all levels of the construction industry is widespread. There are still many countries where it is not widely recognised as a profession. The Services that Building Surveyors undertake are broad but include:

- Construction design and building works
- Project Management and monitoring
- CDM Co-ordinator under the Construction (Design & Management) Regulations 2007
- Property Legislation adviser
- Insurance assessment and claims assistance
- Defect investigation and maintenance adviser
- Building Surveys and measured surveys
- Handling Planning applications
- Building Inspection to ensure compliance with building regulations
- Undertaking pre-acquisition surveys
- Negotiating dilapidations claims.

Building Surveyors also advise on many aspects of construction including:

- design
- maintenance
- repair
- refurbishment
- restoration.

Clients of a building surveyor can be the public sector, Local Authorities, Government Departments as well as private sector organisations and work closely with architects, planners, homeowners and tenants groups. Building Surveyors may also be called to act as an expert witness. It is usual for building surveyors to undertake an accredited degree qualification before undertaking structured training to become a member of a professional organisation. For Chartered Building Surveyors, these courses are accredited by the Royal Institution of Chartered Surveyors. Other professional organisations that have building surveyor members include CIOB, ABE, HKIS and RICS.

With the enlargement of the European community, the profession of the Chartered Building Surveyor is becoming more widely known in other European states, particularly France. Chartered Building Surveyors, where many English speaking people buy second homes.

Distinction from Land Surveyors

In the United States, Canada, the United Kingdom and most Commonwealth countries land surveying is considered to be a distinct profession. Land surveyors have their own professional associations and licencing requirements. The services of a licenced land surveyor are generally required for boundary surveys (to establish the boundaries of a parcel using its legal description) and subdivision plans (a plot or map based on a survey of a parcel of land, with boundary lines drawn inside the larger parcel to indicated the creation of new boundary lines and roads).

Transport Engineering

Transportation engineering is the application of technology and scientific principles to the planning, functional design, operation and management of facilities for any mode of transportation in order to provide for the safe, rapid, comfortable, convenient, economical, and environmentally compatible movement of people and goods (transport). It is a sub-discipline of civil engineering. Transportation engineering is a major component of the civil engineering discipline. The importance of transportation engineering within the civil engineering profession can be judged by the number of divisions in ASCE (American Society of Civil Engineers) that are directly related to transportation. There are six such divisions (Aerospace; Air Transportation; Highway; Pipeline; Waterway, Port, Coastal and Ocean; and Urban Transportation) representing one-third of the total 18 technical divisions within the ASCE (1987).

Figure : *The engineering of this roundabout in Bristol, England, attempts to make traffic flow free-moving.*

The planning aspects of transport engineering relate to urban planning, and involve technical forecasting decisions and political factors. Technical forecasting of passenger travel usually involves an urban transportation planning model, requiring the estimation of trip generation (how many trips for what purpose), trip distribution (destination choice, where is the traveller going), mode choice (what mode is being taken), and route assignment (which streets or routes are being used). More sophisticated forecasting can include other aspects of traveller decisions, including auto ownership, trip chaining (the decision to link individual trips together in a tour) and the choice of residential or business location (known as land use forecasting). Passenger trips are the focus of transport engineering because they often represent the peak of demand on any transportation system.

A review of descriptions of the scope of various committees indicates that while facility planning and design continue to be the core of the transportation engineering field, such areas as operations planning, logistics, network analysis, financing, and policy analysis are also important to civil engineers, particularly to those working in highway and urban transportation. The National Council of Examiners for Engineering and Surveying (NCEES) list online the safety protocols, geometric design requirements, and signal timing.

Transportation engineering, as practised by civil engineers, primarily involves planning, design, construction, maintenance, and operation of transportation facilities. The facilities support air, highway, railroad, pipeline, water, and even space transportation. The design aspects of transport engineering include the sizing of transportation facilities (how many lanes or how much capacity the facility has), determining the materials and thickness used in pavement designing the geometry (vertical and horizontal alignment) of the roadway (or track). Before any planning occurs the Engineer must take what is known as an inventory of the area or if it is appropriate, the previous system in place. This inventory or database must include information on (1) population, (2)l and use, (3) economic activity, (4) transportation facilities and services, (5) travel patterns and volumes, (6) laws and ordinances, (7) regional financial resources, (8) community values and expectations. These inventories help the engineer create business modells to complete accurate forecasts of the future conditions of the system Review.

Operations and management involve traffic engineering, so that vehicles move smoothly on the road or track. Older techniques include signs, signals, markings, and tolling. Newer technologies involve intelligent transportation systems, including advanced traveller information systems (such as variable message signs), advanced traffic control systems (such as ramp metres), and vehicle infrastructure integration. Human factors are an aspect of transport engineering, particularly concerning driver-vehicle interface and user interface of road signs, signals, and markings.

Engineers in this specialisation:

- Handle the planning, design, construction, and operation of highways, roads, and other vehicular facilities as well as their related bicycle and pedestrian realms.
- Estimate the transportation needs of the public and then secure the funding for the project.
- Analyse locations of high traffic volumes and high collisions for safety and capacity.
- Use civil engineering principles to improve the transportation system.
- Utilises the three design controls which are the drivers, the vehicles, and the roadways themselves.

Railroad Engineering

Railway engineers handle the design, construction, and operation of railroads and mass transit systems that use a fixed guideway (such as light rail or even monorails). Typical tasks would include determining horizontal and vertical alignment design, station location and design, and construction cost estimating. Railroad engineers can also move into the specialised field of train dispatching which focuses on train movement control.

Railway engineers also work to build a cleaner and safer transportation network by reinvesting and revitalising the rail system to meet future demands. In the United States, railway engineers work with elected officials in Washington, D.C. on rail transportation issues to make sure that the rail system meets the country's transportation needs.

Port and Harbour Engineering

Port and harbour engineers handle the design, construction, and operation of ports, harbours, canals, and other maritime facilities. This is not to be confused with marine engineering.

Airport Engineering

Airport engineers design and construct airports. Airport engineers must account for the impacts and demands of aircraft in their design of airport facilities. These engineers must use the analysis of predominant wind direction to determine runway orientation, determine the size of runway border and safety areas, different wing tip to wing tip clearances for all gates and must designate the clear zones in the entire port.

Municipal or Urban Engineering

Municipal or urban engineering applies the tools of science, art and engineering in an urban environment. Municipal engineering is concerned with municipal infrastructure. This involves specifying, designing, constructing, and maintaining streets, sidewalks, water supply networks, sewers, street lighting, municipal solid waste management and disposal, storage depots for various bulk materials used for maintenance and public works (salt, sand, etc.), public parks and bicycle paths. In the case of underground utility networks, it may also include the civil portion (conduits and access chambers) of the local distribution networks of electrical and telecommunications services. It can also include the optimising of garbage collection and bus service networks. Some of these disciplines overlap with other civil engineering specialities, however municipal engineering focuses on the coordination of these infrastructure networks and services, as they are often built simultaneously (for a given street or development project), and managed by the same municipal authority.

History

Modern municipal engineering finds its origins in the 19th century United Kingdom, following the Industrial Revolution and the growth of large industrial cities. The threat to urban populations from epidemics of waterborne diseases such as cholera and typhus lead to the development of a profession devoted to "sanitary science" that later became "municipal engineering".

A key figure of the so-called "public health movement" was Edwin Chadwick, author of the parliamentary report "The Sanitary Condition of the Labouring Population", published in 1842.

Early British legislation included:

- Burgh Police Act 1833 - powers of paving, lighting, cleansing, watching, supplying with water and improving their communities.
- Municipal Corporations Act 1835
- Public Health Act 1866 – formation of drainage boards

- Public Health Act 1875 known at the time as the Great Public Health Act.

This legislation provided local authorities with powers to undertake municipal engineering projects and to appoint borough surveyors (later known as "municipal engineers").

In the U.K, the Association of Municipal Engineers, (subsequently named Institution of Municipal Engineers), was established in 1874 under the encouragement of the Institution of Civil Engineers, to address the issue of the application of sanitary science. By the early 20th century, Municipal Engineering had become a broad discipline embracing many of the responsibilities undertaken by local authorities, including roads, drainage, flood control, coastal engineering, public health, waste management, street cleaning, water supply, sewers, waste water treatment, crematoria, public baths, slum clearance, town planning, public housing, energy supply, parks, leisure facilities, libraries, town halls and other municipal buildings.

In the UK, the development of different strands of knowledge necessary for the management of municipal infrastructure led to the emergence of separate specialised institutions, including:

- For drainage: Chartered Institution of Water and Environmental Management, 1895
- For town planning: Town Planning Institute 1914 ... subsequently becoming the Royal Town Planning Institute
- For street lighting: Association of Public Lighting Engineers, 1934 subsequently becoming the Institution of Lighting Engineers
- For highway engineering: Institution of Highways and Transportation, 1930
- For public housing: Institute of Housing, 1931.

In 1984 the Institution of Municipal Engineers merged with the Institution of Civil Engineers. Since the 1970's, there has been a global trend toward increasing privatisation and outsourcing of municipal engineering services. In the UK in the 1990's a change in management philosophy brought the demise of the traditional organisational structure of boroughs where the three functions of town clerk, borough treasurer and borough engineer were replaced by an administrative structure with a larger number of specialised departments. In the late 1990's and early 21st century there was increasing dissatisfaction over what was perceived to be fractured and dysfunctional public services designed along narrow specialities. A more holistic approach to urban engineering began to emerge as an alternative concept. Critics of the specialised approach included the Commission for Architecture and the Built Environment that complained that the specialised approach to management of the public realm focused too much on the on efficient movement of vehicles rather than the more general interests of local communities.

Professional Practice

In the United Kingdom there is no longer any formal professional qualification in municipal engineering although there are degree courses available in urban engineering. A professional certificate in Urban Engineering is available from through the Institution of Incorporated Engineers via the Public Realm Information and Advice Network.

The British Institution of Civil Engineers (ICE) caters to practitioners employed in the public sector, private consultancy and academia through its Proceedings Journal Municipal Engineer. The journal, first published in 1873, has a global scope and covers the whole life cycle of municipal services addressing technical, political and community issues. In addition an Expert Panel responds on behalf of ICE to Government consultations and is represented on the International Federation of Municipal Engineering.

International Organisation

The International Federation of Municipal Engineering (IFME) is an organisation comprising professional municipal engineers from all round the world. IFME's mission is to connect municipal engineers, public works professionals, public agencies, institutions and businesses around the world in order that they can share a global pool of knowledge and experience. The aim is to foster continued improvement in the quality of public works and wider community services.

The inaugural meeting was held in 1960 at the UNESCO headquarters in Paris. Membership has grown steadily and in 2009 comprised representatives from national associations in: Australia, Canada, Denmark, Estonia, Finland, Italy, Israel, The Netherlands, New Zealand, Norway Southern Africa, Sweden and the UK. Belgium and San Marino are presently Corresponding Members.

Related Engineering Disciplines

Municipal or urban engineering combines elements of environmental engineering, water resources engineering and transport engineering.

Relationship to Urban Design or Urban Planning

Today, municipal engineering may be confused with urban design or urban planning. Whereas the urbanist or urban planner may design the general layout of streets and public places, the municipal engineer is concerned with the detailed design. For example, in the case of the design of a new street, the urbanist may specify the general layout of the street, including landscaping, surface finishings and urban accessories, but the municipal engineer will prepare the detailed plans and specifications for the roads, sidewalks, municipal services and street lighting. However Municipal Engineering as practised a century ago fully embraced the function of urban design and urban planning, even though the terms had yet to be coined.

Traffic Engineering (Transportation)

Traffic engineering is a branch of civil engineering that uses engineering techniques to achieve the safe and efficient movement of people and goods on roadways. It focuses mainly on research and construction of the infrastructure necessary for safe and efficient traffic flow, such as road geometry, sidewalks and crosswalks, segregated cycle facilities, shared lane marking, traffic signs, road surface markings and traffic lights.

Traffic engineering is closely associated with other disciplines:

- Transport engineering
- Pavement engineering
- Bicycle transportation engineering
- Highway engineering
- Transportation planning
- Urban planning
- Human factors engineering.

Typical traffic engineering projects involve designing traffic control device installations and modifications, including traffic signals, signs, and pavement markings. However, traffic engineers also consider traffic safety by investigating locations with high crash rates and developing countermeasures to reduce crashes. Traffic flow management can be short-term (preparing construction traffic control plans, including detour plans for pedestrian and vehicular traffic) or long-term (estimating the impacts of proposed commercial developments on traffic patterns). Increasingly, traffic problems are being addressed by developing systems for intelligent transportation systems, often in conjunction with other engineering disciplines, such as computer engineering and electrical engineering.

Traffic Systems

Traditionally, road improvements have consisted mainly of building additional infrastructure. However, dynamic elements are now being introduced into road traffic management. Dynamic elements have long been used in rail transport. These include sensors to measure traffic flows and automatic, interconnected, guidance systems to manage traffic (for example, traffic signs which open a lane in different directions depending on the time of day). Also, traffic flow and speed sensors are used to detect problems and alert operators, so that the cause of the congestion can be determined, and measures can be taken to minimize delays. These systems are collectively called intelligent transportation systems.

Lane Flow Equation

The relationship between lane flow (Q, vehicles per hour), maximum speed (V, kilometres per hour) and density (K, vehicles per kilometre) is:

$$Q = KV$$

Observation on limited access facilities suggests that up to a maximum flow, speed does not decline while density increases. However, above a critical threshold, increased density reduces speed. Additionally, beyond a further threshold, increased density reduces flow as well. Therefore, speeds and lane flows at bottlenecks can be kept high during peak periods by managing traffic density using devices that limit the rate at which vehicles can enter the highway. Ramp metres, signals on entrance ramps that control the rate at which vehicles are allowed to enter the mainline facility, provide this function (at the expense of increased delay for those waiting at the ramps).

Highway Safety

Highway safety engineering is a branch of traffic engineering that deals with reducing the frequency and severity of crashes. It uses physics and vehicle dynamics, as well as road user psychology and human factors engineering, to reduce the influence of factors that contribute to crashes.

A typical traffic safety investigation follows these steps

1. Identify and priorities investigation locations. Locations are selected by looking for sites with higher than average crash rates, and to address citizen complaints.

2. Gather data. This includes obtaining police reports of crashes, observing road user behaviour, and collecting information on traffic signs, road surface markings, traffic lights and road geometry.
3. Analyse data. Look for collisions patterns or road conditions that may be contributing to the problem.
4. Identify possible countermeasures to reduce the severity or frequency of crashes.
    - Evaluate cost/benefit ratios of the alternatives
    - Consider whether a proposed improvement will solve the problem, or cause "crash migration." For example, preventing left turns at one intersection may eliminate left turn crashes at that location, only to increase them a block away.
    - Are any disadvantages of proposed improvements likely to be than the problem you are trying to solve?
5. Implement improvements.
6. Evaluate results. Usually, this occurs some time after the implementation. Have the severity and frequency of crashes been reduced to an acceptable level? If not, return to step 2.

Bicycle Transportation Engineering

Bicycle transportation engineering is the study of transportation engineering as it affects bicycles and cycling. It includes the design of dedicated transport facilities for cyclists, but also the study of how mixed-mode environments (i.e. where cyclists are not segregated from other traffic) can be made to work safely.

Roads

Various methods of altering or reallocating the roadway right-of-way to facilitate bicycling and create bikeways have been added to many of the manuals used by transport planners and engineers.

Segregated Cycle Facilities

Sidepath or Shared use Path: A shared use path is a bikeway that is physically separated from motorised vehicle traffic by an open space (generally landscaping) or some form of barrier.

Cycle Track

A cycle track, also called a *separated cycle lane*, is a bicycle exclusive facility that provide physical separation from motorised vehicle traffic within the road right-of-way, generally by means of kerbs or other barriers. Cycle tracks can either incorporate bicycle-only signal phases at intersections (for 100% separation) or utilise "mixing zones" to merge bicycle and motor vehicle traffic. A cycle track combines the user experience of a separated path with the on-road infrastructure of a bike lane.

Bicycle Lane

A designated bicycle lane, according to the 1998 United States Manual on Uniform Traffic Control Devices, is:

- a portion of a roadway which has been designated for use by bicyclists
- marked by a white (usually solid) stripe painted on the pavement
- significantly narrower than traffic lanes
- found at the side of the traffic lanes.

Cycle lanes require less space compared with the option of providing side paths or separate tracks, and generally do not have the legal / priority issues that facilities separated from the general traffic may have (depending on the locally applicable road laws). However, if traffic is heavy or fast, the lack of physical separation from motor vehicles may make cycle lanes unattractive to less confident cyclists. The lack of separation also means they are more apt to be blocked by (illegally) parked vehicles.

Shoulder

A hard shoulder, or simply shoulder, is a reserved area outside of a roadway, but within the road right-of-way. In addition to (a) serving as a buffer area in the event of a temporary loss of vehicle control, (b) providing emergency access for ambulances and police cars and (c) providing a space for inoperable vehicles, it may also be used by pedestrians and bicyclists.

Non-segregated Facilities

Wide Outside Lane (WOL): Wide outside through lanes help motorists pass slower cyclists without having to decrease speed or change lanes. This method is held by some to be particularly important on roads with a high proportion of wide vehicles, such as buses or heavy trucks. These lanes also provide more room for cyclists to filter past queues of other vehicles in congested conditions. The use of such lanes is specifically endorsed by *Cycling: the way ahead for towns and cities*, the European Commission policy document on cycle promotion.

Cycle friendly infrastructure argues for a marked lane width of 4.25 m (14 ft). It is argued that, on undivided roads, this width provides cyclists with adequate clearance from passing wide vehicles while being sufficiently narrow to deter motorists from attempting to "double up" and form two lanes. This "doubling up" effect may be related to junctions. At non-junction locations, greater width might be preferable if this effect can be avoided.

Shared Bus Lane

A shared bus lane is a bus lane that allows cyclists to use it. Depending on the width of the lane, the speeds and number of buses, and other local factors, the safety and popularity of this arrangement vary. Guidance produced for Cycling England endorses bus lanes as providing cyclists with a *direct and barrier free route into town centres* and as avoiding the difficulties associated with other provisions such as shared-use footways. According to a French survey 42% of cyclists described themselves as "enthusiasts" for shared bus bike lanes versus 33% who were of mixed opinion and 27% who were opposed. Many cycling activists view these as being more attractive than cycle paths, while others object to being in close proximity to bus exhausts. The sharing of bus lanes has been described as "generally very popular" with cyclists in London.

As of 2003, mixed bus/cycle lanes accounted for 118 km of the 260 km of cycling facilities in Paris. The French city of Bordeaux has 40 km of shared bus cycle lanes. It is reported that that in the city of Bristol, a showcase bus priority corridor, where road space was reallocated along a 14 km stretch also resulted in more space for cyclists and had the effect of increasing cycling. The reverse effect has also been suggested, a review carried out in London reports that cycling levels fell across Kew bridge following the removal of a bus lane - this was despite a general increase in cycling level in the city generally. In addition, it is arguably easier, politically speaking, to argue for funding of joint facilities rather than the additional expense of both segregated cycling facilities and bus-only lanes.

In some instances bus lane proposals have run into vehement opposition from cycling representatives - a typical theme is the perceived generation of conflict due to the narrowing of other lanes already shared by cars/cyclists so as to create space for the bus lanes The TRL reports that cyclists and bus drivers tend to have low opinions of each other There have been reports in Dublin of conflict as cyclists choose to cycle in the bus lanes and a bus driver apparently expected them to use adjacent cycle tracks instead. In some other cities the arrangements seem to work successfully, with bus companies and cyclists' groups taking active steps to ensure that understanding is improved between the two groups of road users.

Streets

Shared Space: Shared space schemes, which are characterised by the removal of road markings, signs and signals, give all street users equal priority and equal responsibility for each others safety. Experiences where these schemes are in use, show that street users, particularly motorists, undirected by signs, curbs or road markings, reduce their speed and establish eye contact with other users. Results from the thousands of such implementations worldwide all show casualty reductions and most also show reduced journey times. Following the partial conversion of London's Kensington High Street to *shared space*, accidents were reduced there by 44% (the London average was 17%).

Bicycle Boulevards

A bicycle boulevard is a low speed street which has been designed to discourage cut-through motor vehicle traffic and to give priority to cyclists as through-going traffic.

Bike Paths

Bike paths are bicycle routes that follow an independent right-of-way. There are two distinct types of bike paths: those used exclusively by bicycles and those shared with pedestrians. In the United States almost all bike paths are shared with pedestrians.

Bike paths are often built on old railroad right-of-ways (called rail trails) and as part of greenways and fore shoreways.

Traffic Lights

How traffic signals are designed and implemented directly impacts cyclists. For instance, poorly adjusted vehicle detector systems, used to trigger signal changes, may not correctly detect cyclists. Traffic managers in Copenhagen link cyclist-specific traffic signals on a major arterial bike lane to provide green waves for rush hour cycle-traffic. Cycling-specific measures that can be applied at traffic signals include the use of advanced stop lines and/or bypasses.

The frequency with which lights change is important to cyclists who may conserve energy by anticipating green lights ahead i.e. the shorter the interval the better for cyclists. Intersection clearance times and green wave timing may also be relevant.

Road Surface

Bicycle tires being narrow, road surface is more important than for other transport, for both comfort and safety. The type and placement of storm drains, manholes, surface markings, and the general road surface quality should all be taken into account by a bicycle transportation engineer. Drain grates, for example, must not catch wheels.

Parking

Bicycle parking is another important part of Bicycle Transportation Engineering. In most of the United States, bicycle parking facilities are scarce, or are so inadequate that nearby trees or parking metres are used. The hitching post type of bicycle stand is an improvement over the old type that had a slot for the front wheel but only allow for two bicycles per post. The Netherlands, where bicycles are much in use, has two-tiered bicycle racks giving high density (the handlebars overlap) and security (the bicycle is held well and is easy to lock).

Secure bicycle parking is argued to be a key factor influencing the decision to cycle. To be considered secure, the parking must be of a suitable design: allowing the bicycle to be locked via the frame. A readily observable location can also permit so-called passive security from passers-by. Weather protection is also desirable. As a rule, where cycling is encouraged as an alternative to motoring, efforts are made to make bicycle parking more convenient and attractive to use than nearby car parking arrangements. This usually means providing a wide distribution of visible, clearly designated parking spots, close to the entrances of destinations being served.

Storage rooms or bicycle lockers may also be provided. In some cases large concentrations of bike parking may be more appropriate, sometimes being supervised and sometimes charging a fee. Examples include large bike parking facilities at public transport interchanges such as railway, subway, tram, bus stations, or at bike stations where they may be useful in Mixed-mode commuting.

Conversely, where cycling is seen as an unwelcome or inappropriate activity, or due to lack of knowledge about best practices, bicycle parking may simply not be provided or else placed at awkward, distant, and out-of-sight locations. Cyclists may be expressly forbidden from parking their bicycles at the most convenient locations. In April 2007, the authorities at the University of California's Santa Barbara campus started confiscating bicycles not parked at the allegedly inconvenient official bike racks. Often, property owners display signage on fencing to discourage bicyclists from locking their bicycles.

Pavement Engineering

Pavement engineering is a branch of civil engineering that uses engineering techniques to design and maintain flexible (asphalt) and rigid (concrete) pavements.

This includes streets and highways and involves knowledge of soils, hydraulics, and material properties. Pavement engineering involves new construction as well as rehabilitation and maintenance of existing pavements. Maintenance often involves using engineering judgment to make maintenance repairs with the highest long-term benefit and lowest cost. The Pavement Condition Index (PCI) is an example of an engineering approach applied to existing pavements. Another example is the use of a falling weight deflectometre (FWD) to non-destructively test existing pavements. Calculation of pavement layer strengths can be performed from the resulting deflection data.

Four on the Floor! (Strength vs. Stability)

There are two issues that will be emphasized throughout this course that are crucial to the understanding of Architectonics:

- Strength : the capacity of the individual elements, which together make up a structural system, to withstand the load that is applied to it.
- Stability: the capability of a structural system to transmit various loadings safely to the ground.

These two critical issues are experienced daily from the moment that an individual is born. A newborn baby cannot even hold its own head upright. The large mass of the head requires a support system that has sufficient strength to enable the head to maintain its stability. This steadily increases as the bones, muscles and tendons of the skeletal and muscular systems increase in strength. Eventually the extra support provided by the arm or hand is no longer needed. The first challenge posed by gravity is overcome.

Crawling on four points of support proves to be a very stabile situation for quite a long time. The "leap" to the unstable two point stance is the next development in our understanding of the influence of gravity. Again, the structural system must develop to the point that the individual elements of the system have acquired sufficient strength. The first steps are made: an action of supreme coordination of hundreds of elements that becomes second nature to homo sapiens. The list can be extrapolated to touch on many aspects of the human experience; riding tricycles and bicycles, jumping on trampolines, exercising on parallel bars, sliding on ice skates, sailing in a heavy wind, rocking a small boat, the list is endless. These are part of the human experience and each and every one rely on an inherent understanding of strength and stability.

How many times has a parent scolded a child to "put four on the floor!!!"? What the parent really means to say is, "if you do not put all of the legs of your chair on the ground, you are going to tip over!" Both strength and stability issues are addressed in this simple exclamation. Under normal conditions, the elements which make up the chair (its legs, bracing and seat) can easily resist the implied vertical loads. The strength of the individual elements of the chair have been designed for this type of static load. The seat (as a horizontal load-bearing element) must transfer its load through a connection to the legs (vertical load-bearing elements). Granted, some chairs will withstand a greater load than others, but they all resist the pull of gravity on the person sitting in them. If the legs cannot support the applied load they will fracture or break. These are examples of strength failure.

The stability of the system of elements depends upon the orientation of the chair in space. When it stands upright, on all four legs, it is a stable system. If it is on it's side, the chair might not be able to resist the loads for which it was designed. As it is tilted onto the back two legs, the structural system loses its equilibrium. At a certain point the chair as a system becomes unstable, fails and gravity pulls the supported load to the ground. This is a stability failure. In this type of failure, the individual elements retain their strength even as the system fails. The chair (system) could also have failed if the two supporting legs had experienced a strength failure (broken).

In each of these situations the chair, as a structural system, has reached the limit of its strength. As the saying goes, a chain (structural system) is only as strong as the weakest link (element)!

Any structural system can be studied in light of these two issues. For example, the column of the Greek temple shown above is an element that can experience a strength (crushing) failure, or a system (buckling) failure. It is/was part of a larger structural system.

The Role of Structural Failure

The concepts of element strength and system stability are learned at an early age. They often appear in children's stories that have been repeated for decades. One favourite in North America is the story of the "Three Little Pigs." Each pig desired to build a house to protect itself from the Big, Bad Wolf. One pig built a house of straw while another built a house of sticks and twigs. The third pig erected a house of brick. The story follows that the big, bad wolf came to the homes of the three pigs. He drew a huge breath and huffed and puffed until he blew the first house down. He chased the pig into the second house where he repeated his imitation of a typhoon. This house was also blown over and the pigs fled into the brick house. In front of the brick house the wolf hyperventilated. His grandest attempts to blow over this ediface failed and so he did not have pork for dinner. The story teaches children that a brick building is the strongest and most stable becuase it is the only one which could resist the huffing and puffing. The straw and stick buildings were blown away in the wind and only within the brick structure do the three pigs find their security. This lesson is correct if the load that is of greatest concern is a tornado, typhoon, hurricane or similar lateral load. However, the brick house would have been the worst structure to live in if it was built in an earthquake zone. In fact, an unreinforced masonry building is much more life threatening in an earthquake when compared to a straw hut. Then again, another response would be appropriate if the house was in a region where winds were not a real problem, but extreme temperature variations were critical. The same would be true if mobility was the most significant issue. Each of these are specific loading conditions that could lead to structural failure. It is critical that the appropriate loads and loading combinations are considered for each and every structure depending on use and location. Every one is different!

"Humpy Dumpty" is another childrens' tale with a structural lesson. He was a good egg who sat on a wall for quite a while. One day he lost his balance and fell to the ground where all of the king's men could not put Humpty back together again. The egg shell is a wonderfully efficient, naturally occurring thin structure that is extremly strong under evenly distributed loading conditions. One type of load which a thin shell cannot resist is a sharp point load. Thus, when Humpty fell to the ground he loaded his wonderful structure in a way that caused a strength and system failure. Building Codes are often considered to be the bane of architectural designers. However, they have been a part of the profession for thousands of years. The illustration above is a part of the Code of Hammurabi that was written about 2200BC. The Code can be translated as:

If a builder builds a house for a man and does not make its construction firm and the house collapses and causes the death of the owner of the house - that builder shall be put to death. If it causes the death of a son of the owner - a son of that builder shall be put to death.

If it causes the death of a slave of the owner - the builder shall give the owner a slave of equal value. If it destroys property - the builder shall restore whatever it destroyed and because the builder did not make the house firm, shall rebuild the house which collapsed at his own expense.

If a builder builds a house and does not make its construction meet the requirements and a wall falls in - that builder shall strengthen the wall at his own expense.

The negative incentive of the Code of Hammurabi does not bode well for innovation. However, it clearly demonstrates the point that building codes have existed for a long time, and are not likely to go away. Structural designers are often considered to be a very conservative group. One of the reasons for this could be found in such a building code. In addition, when a building experiences structural failure the results are usually quite different than if the air conditioning system fails. Although some aspects of the building codes seem ridiculous, they are in place to guarantee the public's safety and should be respected.

Each and every one of us is a structural designer in their own way. The act of walking down a set of stairs or pushing a wheelchair forward or swimming across a pool requires an understanding of gravity, the flow of forces and how to work with them.

Petroski notes: "The Concept of Failure is central to understanding engineering, it has as its first and foremost objective the obviation of failure. Design, even structural design, is a human endeavour and thus it is subject to error. Due to this, some designs are destined to fail. This can lead to a loss of life which in itself is tragic, but a deeper unforgivable tragedy exists when the lessons of the failure are understood and allowed to occur again."

What is a Force?

The one constant around the world is the action of gravity upon each and every structure that is erected. The primary function of all structural design is to make a building stand-up. Understanding architectonics will enable a designer to include these issues as part of a design language that will create a significantly clearer architectural expression. Primary to this study is the concept of a force. A force is actually a very abstract conception. It can be defined, but it cannot become physically apparent until it meets resistance. Imagine a six foot tall block of ice sliding along a frictionless surface laid inside of a hockey rink. If a person props herself against the wall and tries to stop the ice she will then perceive the force imparted by the block of ice. She transfers the force of the block of ice into a force that moves into the ground. Thus, if the ground could actually experience a push, it would as well as the slide is arrested.

A "force" is an action that changes, or tends to change, the state of motion of the body upon which it acts. It is a vector quantity that can can be represented either mathematically or graphically. A complete description of a force MUST include its:

1. Magnitude
2. Direction qnd Sense
3. Point of Application.

The Magnitude is most often expressed in the units of pounds (lbs), newtons (N), kilo-newtons (KN) or kilo-pounds (KIPS; 1 Kip = 1000 lbs). The magnitude is represented graphically by the scaled length of the arrow which represents the force. Graphic statics depends upon the accurate representation of the magnitude of each force acting upon a body.

The Direction of a force is discerned by oberserving the line of action of the vector of the force. The Sense refers to the direction of the movement of the vector along that line of action.

The sense is always represented on the vector by an arrowhead. These two attributes can be either a written verbal description or more conveniently expressed in terms of 360 degrees. In the later case, one begins with zero and increases clockwise with the direction of the arrowhead until 360 is reached.

The sense of the force can also be expressed as a positive or negative sign. This is especially useful when combining forces algebraically.

It is very important not to confuse the direction and sense. The direction always relates to the line of action of the vector, and the sense is the way in which the vector would move along that line. In the example, the sense of the 100 lb force would be "down and to the left" or "210 degrees."

The Point of Application is often overlooked in the description of a force. However, it is fully as important as the magnitude of the force! It is the exact location of the application of a force on a body. It can either be a relative measurement or a set of coordinates.

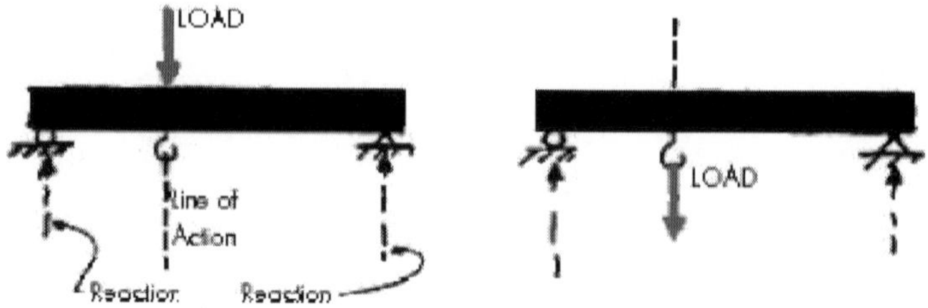

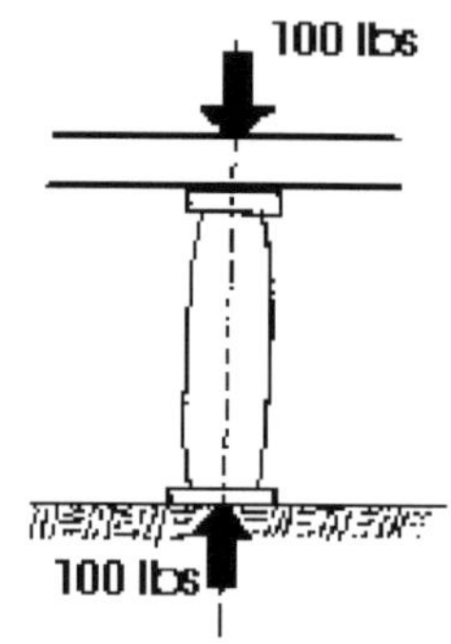

A 100 kip force is applied to the stone column in the diagram. The column will experience this force at every point along the line of action of the force. As a matter of fact, the force will also be transferred to the ground which is supporting the column. Thus, the earth below the column along the line of action of the force will also experience the 100 kip load. This illustrates the Principle of Transmissibility. The point of application of an external force acting on a body (structure) may be transmitted anywhere along the force's line of action without affecting the other external forces (reactions and loads) acting on that body. This means that there is NO NET CHANGE in the static effect upon any body if the body is in equilibrium. This can be illustrated with the following diagram.

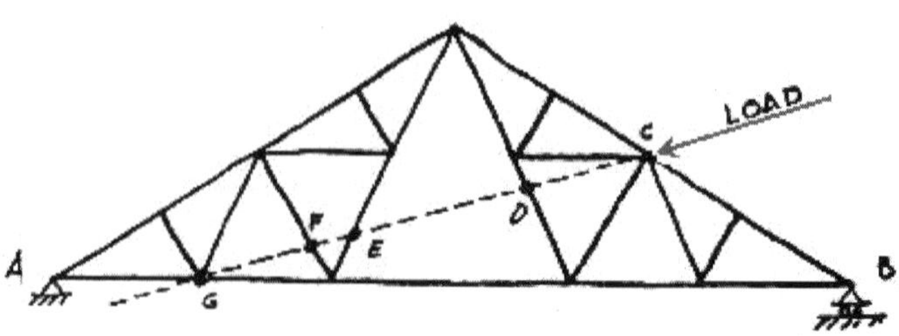

Assume that a beam is supported at its ends. A load is applied to the top of the beam that is acting downward. This load could be a person standing on the beam. The load creates reactions that push up at the two points of support. The line of action of the load (person) on the beam also passes through a hook that is attached to the underside of the beam. Now, if the person standing on top of the beam would climb down and hold on to the hook exactly below the point where they were previously located so that the lines of action were exactly the same, the reactions at the ends of the beams would not change. This is because the load of the person is still acting along the same line of action. As long as a load is applied at any point along the line of action the external reactions will not change.

The truss above is loaded with a force that is applied at point C. This load creates reactions at the two supports A and B. The load on the truss could move anywhere along the line of action and the external reactions at A and B would remain the same. That means that if the load was applied at points D, E, F or G the reactions at A and B would not change.

Note that the only point of discussion at this moment is the fact that the external reactions will not change. It is clear that the internal forces will vary greatly within the truss as the force is moved along the line of action. This issue will be discussed in a later lecture.

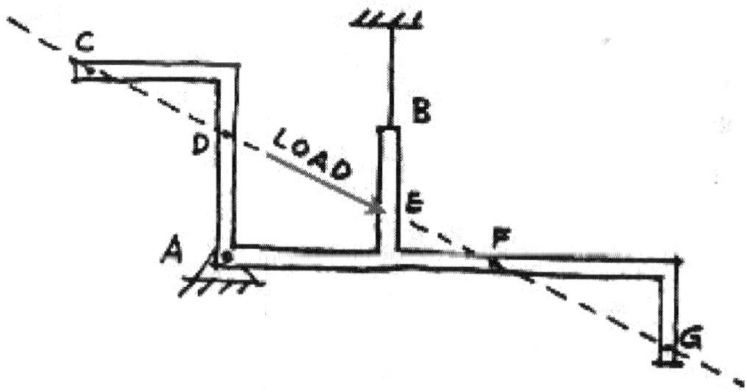

This strangely shaped beam is another example of a structure that is loaded at a specific point, namely E. In order for the structure to remain at rest there must be reactions of some kind at points A and B. The reaction at B is a tension reaction since it is a cable. Again, if the load was applied at points C, D, E, F or G on this rigid body (structure) the reactions at A and at B would remain exactly the same.

Only the external forces (reactions) remain unchanged. Some of the internal resisting forces within the elements of the structure change as the load is applied at different points along its line of action. This illustrates one important difference between INTERNAL and EXTERNAL forces.

The Principle of Transmissibility applies to any body (blobs, balloons, simple beams, crooked beams, trusses, shells, etc.). It is independent of the body's size or shape.

Moments

The Moment of a force is a measure of its tendency to cause a body to rotate about a specific point or axis. This is different from the tendency for a body to move, or translate, in the direction of the force. In order for a moment to develop, the force must act upon the body in such a manner that the body would begin to twist. This occurs every time a force is applied so that it does not pass through the centroid of the body. A moment is due to a force not having an equal and opposite force directly along it's line of action.

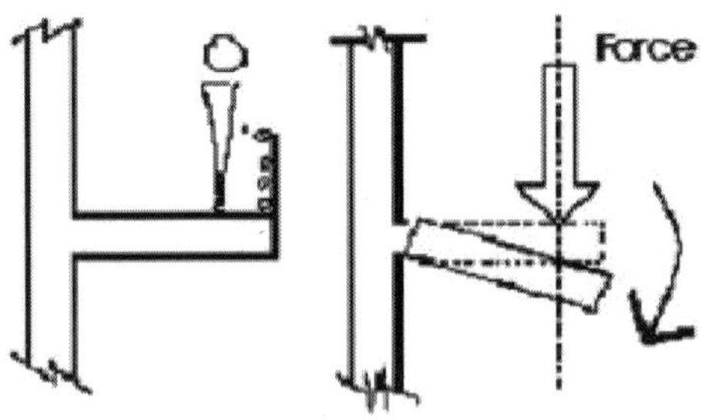

Imagine two people pushing on a door at the doorknob from opposite sides. If both of them are pushing with an equal force then there is a state of equilibrium. If one of them would suddenly jump back from the door, the push of the other person would no longer have any opposition and the door would swing away. The person who was still pushing on the door created a moment.

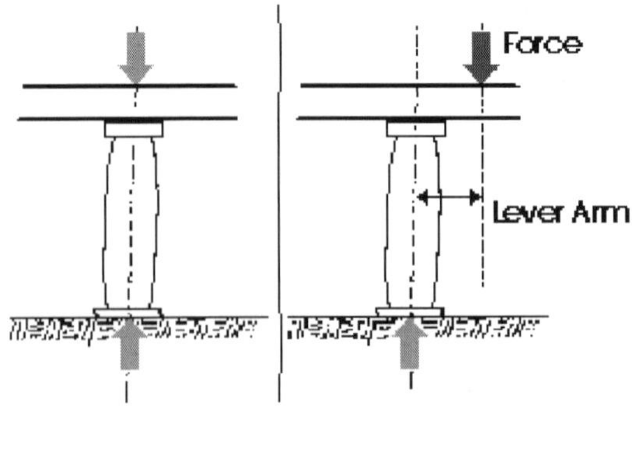

The magnitude of the moment of a force acting about a point or axis is directly proportional to the distance of the force from the point or axis. It is defined as the product of the force (F) and the moment arm (d). The moment arm or lever arm is the perpendicular distance between the line of action of the force and the centre of moments.

The Centre of Moments may be the actual point about which the force causes rotation. It may also be a reference point or axis about which the force may be considered as causing rotation. It does not matter as long as a specific point is always taken as the reference point. The latter case is much more common situation in structural design problems. A moment is expressed in units of foot-pounds, kip-feet, newton-metres, or kilonewton-metres. A moment also has a sense; it is either clockwise or counter-clockwise. The most common way to express a moment is :

$$Moment = Force \times Distance$$

$$M = F \times d$$

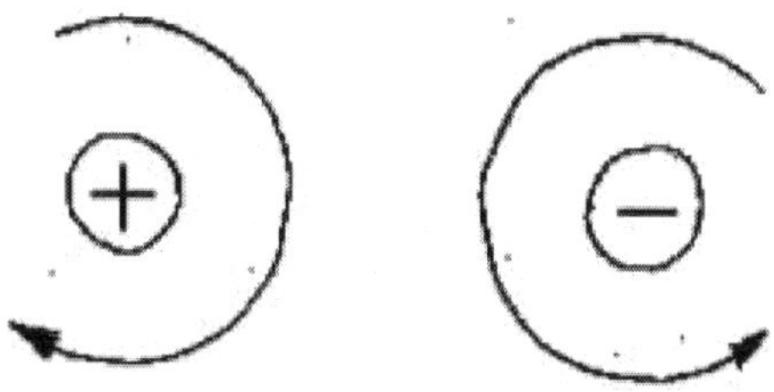

The example shown is a wrench on a nut (at C) that has a force applied to it. The force is applied at a distance of 12 inches from the nut. The centre of moments could be point C or points A or B. The moment arm for calculating the moment around point C is 12 inches. The magnitude of the moment of the 100 pound force about point C is 12 inches multiplied by the force of 100 pounds to give a moment of 1200 inch-pounds (or 100 foot-pounds). Similarly, the moment about point A can be found to be 800 inch-pounds.

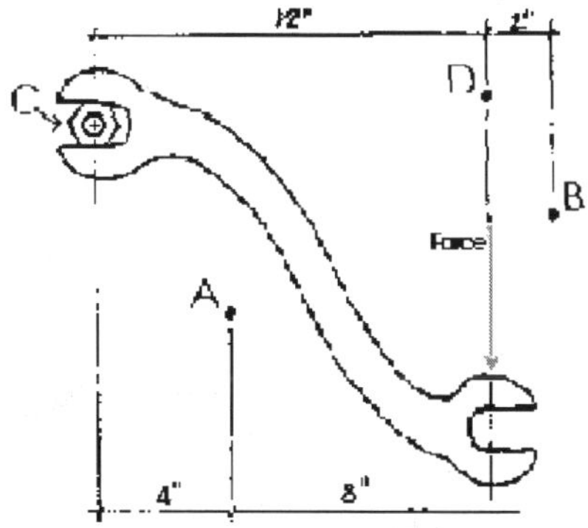

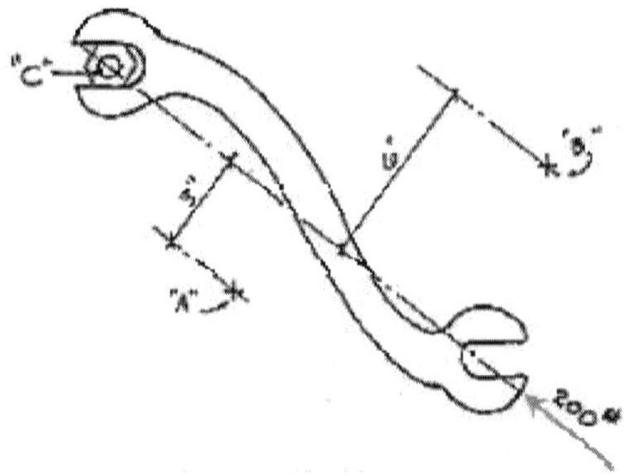

The direction of the rotation is important to understand in order to describe its effect on the body (structure). A moment will cause either a clockwise or counter-clockwise rotation about the centre of moments. It is essential that the direction of rotation about the centre of moments be understood. Since an international convention does not exist, in this course, a clockwise rotation about the centre of moments will be considered as positive; a counter-clockwise rotation about the centre of moments will be considered as negative. A moment causes a rotation about a point or axis. Thus, the moment of a force taken about any point that lies on its own line of action is zero. Such a force cannot cause a rotation because the moment arm is nonexistant.

In this second example, a 200 pound force is applied to the wrench. The moment of the 200 pound force applied at C is zero because 200 pounds x 0 inches = 0 inch-pounds (Fxd = M). In other words, there is no tendency for the 200 pound force to cause the wrench to rotate the nut. One could increase the magnitude of the force until the bolt finally broke off.

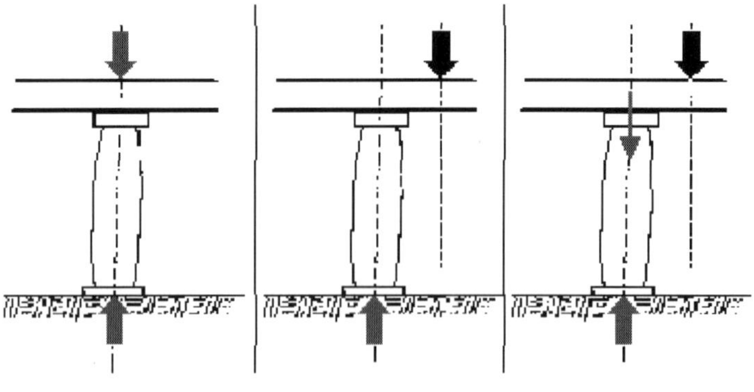

A moment can also be considered to be the result of forces detouring from a direct line drawn between the point of loading of a system and its supports. In this case, the blue force is an eccentric force. In order for it to reach the base of the column, it must make a detour through the beam. The greater the detour, the greater the moment. The most efficient structural systems have the least amount of detours possible. This will be discussed in more detail in Lecture 37 and later courses.

There are cases in which it is easier to calculate the moments of the componenets of a force around a certain point than it is to calculate the moment of the force itself. It could be that the determination of the perpendicular distance of the force is more difficult than determining the perpendicular distance of components of the force. The moment of several forces about a point is simply the algebraic sum of their component moments about the same point. When adding the moments of componenets, one must take great care to be consistent with the sense of each moment. It is often prudent to note the sense next to the moment when undertaking such problems.

Any difficulty with calculating a moment can usually be traced to one of the following:

1. The centre of moments has not been correctly established or clearly understood.
2. The assumed moment arm is not the PERPENDICULAR distance between the line of action of the force and the centre of moments.
3. The direction, or sense, of the rotation has been ignored or misunderstood.

Equilibrium

Architectural structures are normally stationary. Most clients, building officials and designers prefer that their structures remain static rather than move dynamically. There are specific loading conditions which are dynamic loads, but in each and every case a return to a stable and static state is desirable. Such a condition is known as equilibrium.

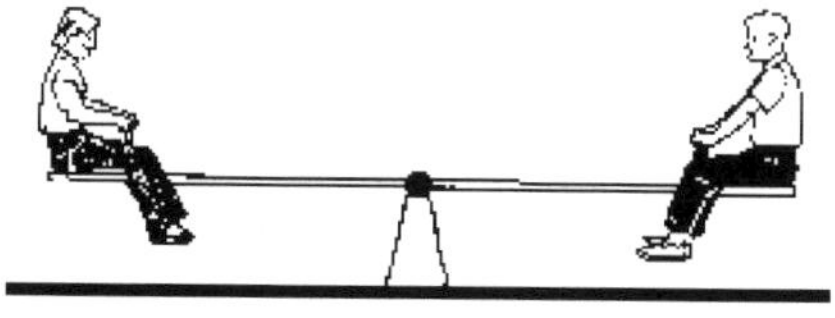

Various states of static equilibrium are experienced throughout one's life. Think of the "teeter-totter" at a playground or of a game of "tug-of-war." In the first case, two or more individuals sit upon a board which has been fixed to a fulcrum which allows rotation. If each of the individuals on the teeter totter weight exactly the same amount and sit at exactly the same distance from the fulcrum the teeter-totter will not move. A state of equilibrium has been achieved. The two will remain at rest until an action takes them out of equilibrium.

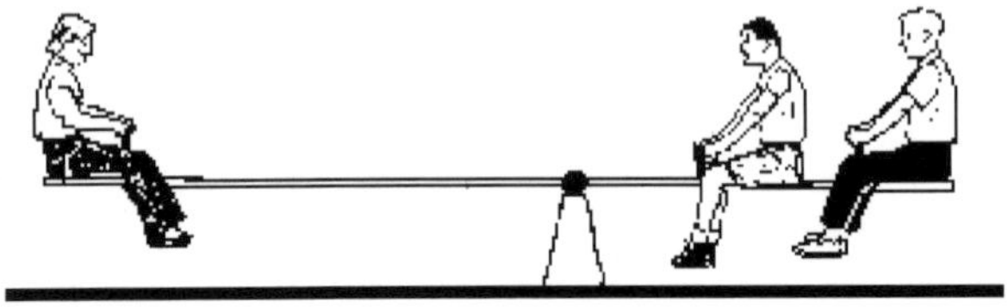

Such an action could be the addition of another person to the system or it could be that one of the original two would change their position slightly. In either case, the teeter-totter would most likely swing to one side and rest upon the ground. A new state of equilibrium would have been found. In this case, another boy climbed on behind the one already sitting on the right. In order to put the system back into equilibrium the girl on the left had to move backwards along the board until she was far away. The moment two boys created around the fulcrum doubled when the second boy climbed on. The girl knew that the only way that she could increase the magnitude of the moment she created would be to increase the moment arm. Thus she moved back until she was twice as far from the fulcrum. Now the system would be back in equilibrium.

Another example of a state of equilibrium is the game of "tug-of-war." A rope is pulled taught between two teams; each pulls with a force that equals the force of the opposite team. Assume in the figure that each team is pulling with a force of 220 pounds. As long as each team maintained a pull of 220 pounds the system is in equilibrium. If during this time a device would be inserted between the two teams to measure the magnitude of the tension force that the rope has anywhere along its length, it would read 220 pounds at each and every point. This would be true at ANY point along the rope. A structure is in equilibrium when all forces or moments acting upon it are balanced. This means that each and every force acting upon a body, or part of the body, is resisted by either another equal and opposite force or set of forces whose net result is zero. Issac Newton addressed this issue when he noted that a body is at rest will remain at rest until acted upon by an external force. Every structure that can be seen to remain standing on a daily basis is in equilibrium; it is at rest and each of its members, combination of its members or any part of a member that is supporting a load are also at rest. There is a net result of zero in all directions for all of the applied loads and reactions.

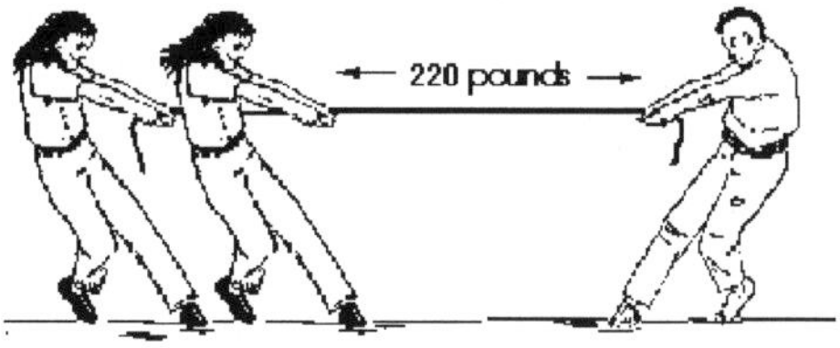

The beam of the top figure illustrates the system at rest under it's own weight. The lower figure shown another state in which a pair of scissors sits at one end and a thermos of coffee on the opposite side near the fulcrum. All of the forces and moments are balanced so that the system is in a stable equilibrium.

There are two types of equilibrium; External and Internal. External equilibrium encompasses the loads upon, and reactions of, a structural system as a whole. Internal equilibrium describes the various forces the are acting within every member of the system. There are conditions of equilibrium that must be satisfied for each case. These are:

Sum of All Vertical Forces (Fy) = 0

Sum of All Horizontal Forces (Fx) = 0

Sum of All Moments (Mz) = 0

(Sum of All Forces (Fz) = 0)

(Sum of All Moments (My) = 0)

(Sum of All Moments (Mx) = 0)

These six equations are all that can be used to determine every one of the forces that are acting with a structure. They are few, but very powerful. The first three are the most common equations and will be utilised in all of the problems associated with third course. The other three are only necessary when considering three-dimensional force systems.

Chapter 4

The Elements of Coplanar Force Resolution

There are many ways in which forces can be manipulated. It is often easier to work with a large, complicated system of forces by reducing it an ever decreasing number of smaller problems. This is called the "resolution" of forces or force systems. This is one way to simplify what may otherwise seem to be an impossible system of forces acting on a body. Certain systems of forces are easier to resolve than others. Coplanar force systems have all the forces acting in in one plane. They may be concurrent, parallel, non-concurrent or non-parallel. All of these systems can be resolved by using graphic statics or algebra.

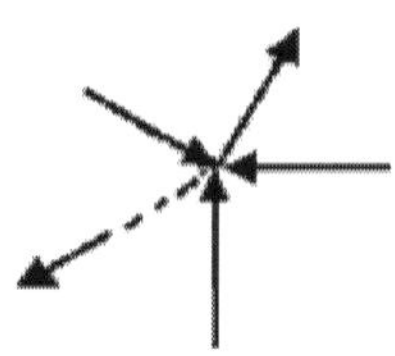

Concurrent

A concurrent coplanar force system is a system of two or more forces whose lines of action ALL intersect at a common point. However, all of the individual vectors might not actually be in contact with the common point. These are the most simple force systems to resolve with any one of many graphical or algebraic options.

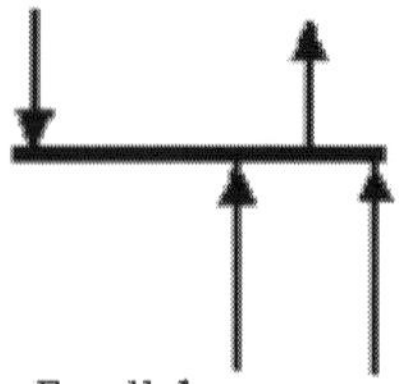

Parallel

A parallel coplanar force system consists of two or more forces whose lines of action are ALL parallel. This is commonly the situation when simple beams are analysed under gravity loads. These can be solved graphically, but are combined most easily using algebraic methods.

non-concurrent and non-parallel

The last illustration is of a non-concurrent and non-parallel system. This consists of a number of vectors that do not meet at a single point and none of them are parallel. These systems are essentially a jumble of forces and take considerable care to resolve.

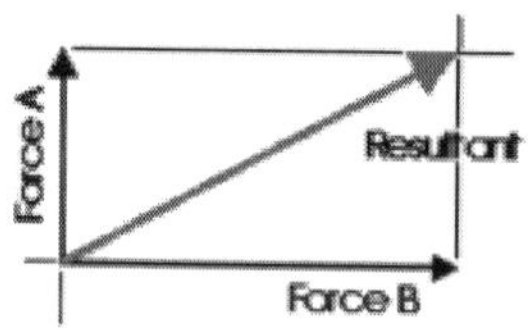

Almost any system of known forces can be resolved into a single force called a resultant force or simply a Resultant. The resultant is a representative force which has the same effect on the body as the group of forces it replaces. (A couple is an exception to this) It, as one single force, can represent any number of forces and is very useful when resolving multiple groups of forces. One can progressively resolve pairs or small groups of forces into resultants. Then another resultant of the resultants can be found and so on until all of the forces have been combined into one force. This is one way to save time with the tedious "bookkeeping" involved with a large number of individual forces. Resultants can be determined both graphically and algebraically. The graphical methods that will be discussed in Lecture 8 are the Parallelogram Method and the Triangle Method. It is important to note that for any given system of forces, there is only one resultant.

It is often convenient to decompose a single force into two distinct forces. These forces, when acting together, have the same external effect on a body as the original force. They are known as components. Finding the components of a force can be viewed as the converse of finding a resultant.

There are an infinant number of components to any single force. And, the correct choice of the pair to represent a force depends upon the most convenient geometry. For simplicity, the most convenietent is often the coordinate axis of a structure.

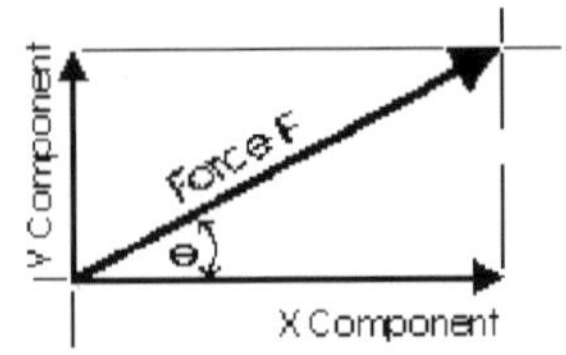

This diagram illustrates a pair of components that correspond with the X and Y axis. These are known as the rectangular components of a force. Rectangular components can be thought of as the two sides of a right angle which are at ninety degrees to each other. The resultant of these components is the hypotenuse of the triangle. The rectangular components for any force can be found with trigonometrical relationships: $F_x = F \cos \theta$, $F_y = F \sin \theta$. There are a few geometric relationships that seem to common in general building practise in North America. These relationships relate to roof pitches, stair pitches, and common slopes or relationships between truss members. Some of these are triangles with sides of ratios of 3-4-5, 1-2-sqrt3, 1-1-sqrt2, 5-12-13 or 8-15-17. Committing the first three to memory will simplify the determination of vector magnitudes when resolving more difficult problems.

When forces are being represented as vectors, it is important to should show a clear distinction between a resultant and its components. The resultant could be shown with colour or as a dashed line and the components as solid lines, or vice versa. NEVER represent the resultant in the same graphic way as its components.

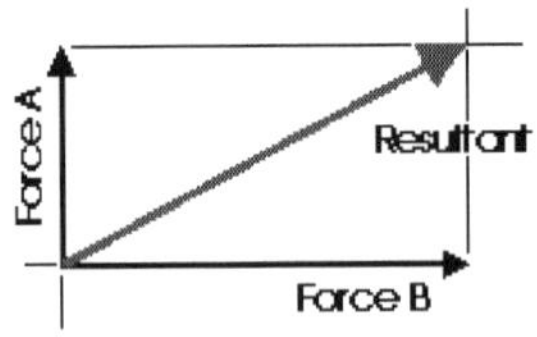

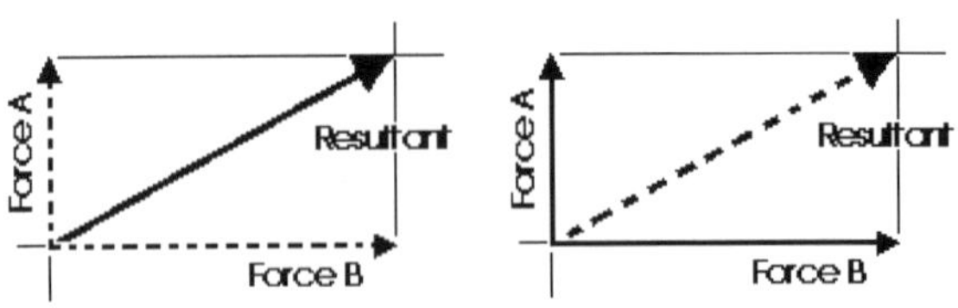

Any concurrent set of forces, not in equilibrium, can be put into a state of equilibrium by a single force. This force is called the

Equilibrant. It is equal in magnitude, opposite in sense and co-linear with the resultant. When this force is added to the force system, the sum of all of the forces is equal to zero. A non-concurrent or a parallel force system can actually be in equilibrium with respect to all of the forces, but not be in equilibrium with respect to moments.

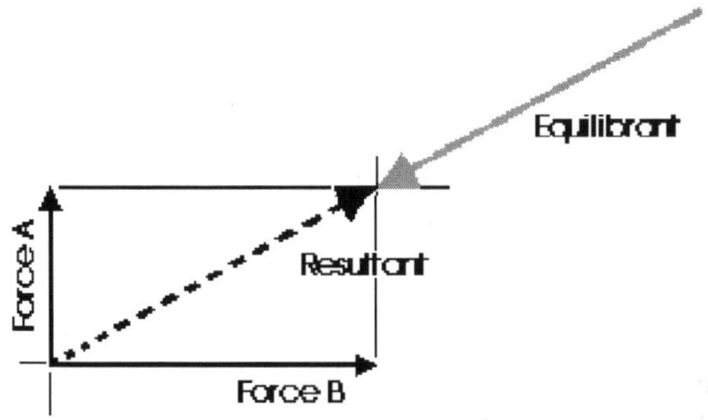

Graphic Statics and graphical methods of force resolution were developed before the turn of the century by Karl Culmann. They were the only methods of structural analysis for many years. These methods can help to develop an intuitive understanding of the action of the forces. Today, the Algebraic Method is considered to be more applicable to structural design. Despite this, graphical methods are a very easy way to get a quick answer for a structural design problem and can aid in the determination of structural form.

Graphic Methods of Coplanar Force Resolution

The Parallelogram of Forces Method is one of the graphical methods developed to find the resultant of a coplanar force system. Two or more concurrent forces can be replaced by a single resultant force that is statically equivalent to these forces.

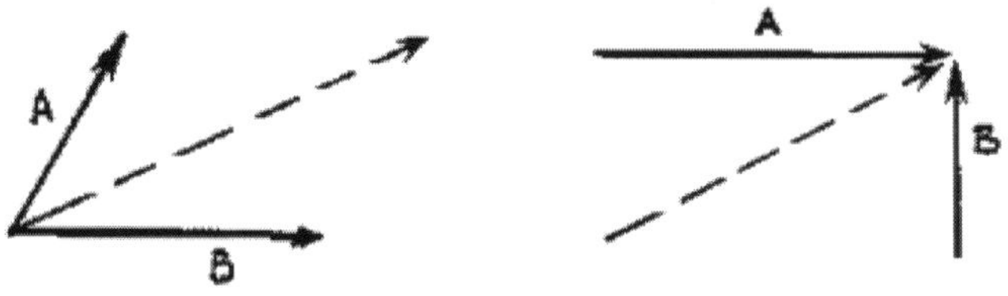

The illustration shows two vectos and their resultant. The resultant force is shown as the dashed vector. In order to resolve these forces graphically, one must first extend the lines of action of two concurrent forces until they intersect. This intersection is known as the point of origin for the system. Both forces, as well as the resultant, must ALL act either away from or toward the point of origin.

The resultant can be represented graphically by the diagonal of the parallelogram formed by using the two force vectors to determine the length of the sides of the parallelogram. The magnitude of the resultant can be accurately measured as the scaled length of the diagonal. The resultant MUST go through the point of intersection of its components!!!

(Remember: graphical solutions depend upon the accuracy of the drawing. The length of each vector should be carefully scaled to equal the magnitude of the force).

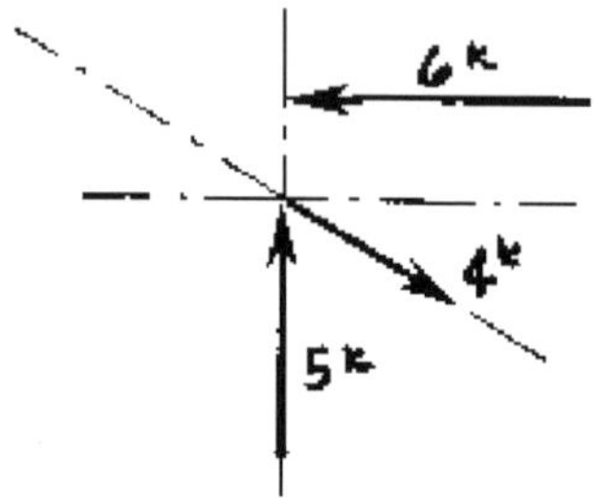

More than two non-parallel forces can be combined by successively eliminating one of the forces. Combine any two of the forces into their resultant by the parallelogram method. Combine this resultant with any of the remaining forces (or with the resultant of any of the remaining forces) until all of the forces are included. One must remember that the vectors can only be translated (or moved) along their lines of action. Two vectors (or Forces) cannot be combined (or resolved) until both of them are meeting head-to-head or tail-to-tail!

The resolution of the this system is a single vector that has a magnitude of approximately 4k with a direction of up and to the left. Try it yourself!!! The Triangle of Forces Method is another graphical method developed to find the resultant of a coplanar force system. Since the opposite sides of a parallelogram are equal, a force triangle may also be found instead of using the parallelogram method. This method is quite useful because it can be simultaneously applied to any number of concurrent forces.

Parallelogram of Forces Method   Triangle of Forces Method

To calculate the resultant of the force system shown above, move force A so that the components A and B form a "Head-to-Tail" arrangement. The resultant R is found by starting at the tail of B (the point of intersection of forces A and B) and terminating at the head of the transposed A. Force B could have been transposed instead of force A. If this had been done, the resultant would have started from the tail of A and terminated at the head of force B.

The resultant is described by the vector's magnitude and direction. These are determined by scaling the length and angle respectively. The accuracy of these values depends upon the accuracy of the graphics.

More than two non-parallel, non-concurrent forces can be combined by successively eliminating one of the forces. Combine any two of the forces into their resultant by the triangle method, and then extend that resultant until it intersects the line of action of another force. One continues this process until all forces have been included. In this way, each one of the forces is successively combined with the resultant of the previous triangle. One cannot simply continue to add the vectors head-to-head or tail-to-tail because the resulting lines of action would then be incorrect!

Algebraic Methods of Coplanar Force Resolution

A force can be resolved into an infinite number of combinations of components by the parallelogram method. The most useful components are the two components that are parallel to the X and Y axes. These are known as the rectangular components.

Random Components Rectangular Components

The solution of many problems is greatly simplified by using the rectangular components. These are usually noted as Fx, for forces parallel to the X axis, and Fy, for forces parallel to the Y axis. They are also known simply as H, for horizontal, and V, for vertical.

The rectangular components may be determined graphically, where the force is shown as a vector, or algebraically.

In order to resolve a vector into its components, $Fx = F \cos ø$ or $Fy = F \sin ø$, one must know at least two items of the six geometric descriptors of a triangle (the lengths of the sides and the three angles).

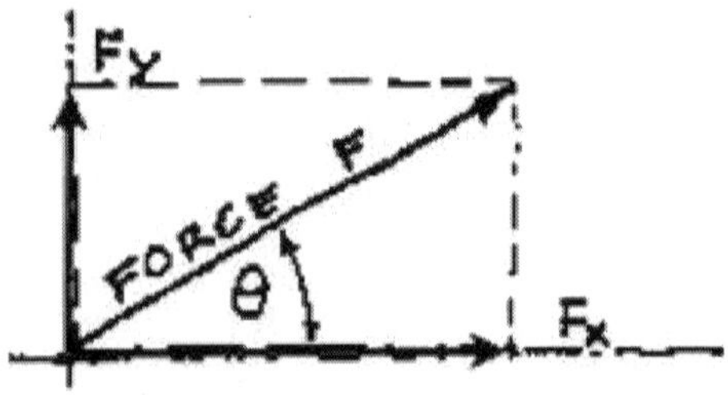

Conversely, if the magnitude of the rectangular components are known, the resultant can be found with the Pythagorean Theorem, F = SQRT (Fx^2 + Fy^2). The direction of the resultant may be determined by trigonometry knowing that the tan ø is = opposite side/ adjacent side.

The solutions to many problems are greatly simplified by resolving multiple forces into a resultant or by finding the rectangular components of a force system. This tool will be used extensively throughout the course.

In summary:

Fx = F cos ø

Fy = F sin ø

F = SQRT (Fx^2 + Fy^2)

tan ø = Fx / Fy = opposite side / adjacent side.

The resultant of any number of concurrent forces can be found by resolving each force into its rectangular components and then adding the components algebraically. Remember, the sum of Fx = 0 and Fy = 0. The resulting numbers will be the components of the resultant. From these, the resultant can be found as before by using the Pythagorean Theorem in which R = SQRT (Fx^2 + Fy^2) and the formula for tan ø = Fy / Fx.

Principle of Moments

The Principle of Moments, also known as Varignon's Theorem, states that the moment of any force is equal to the algebraic sum of the moments of the components of that force. It is a very important principle that is often used in conjunction with the Principle of Transmissibility in order to solve systems of forces that are acting upon and/or within a structure. This concept will be illustrated by calculating the moment around the bolt caused by the 100 pound force at points A, B, C, D, and E in the illustration.

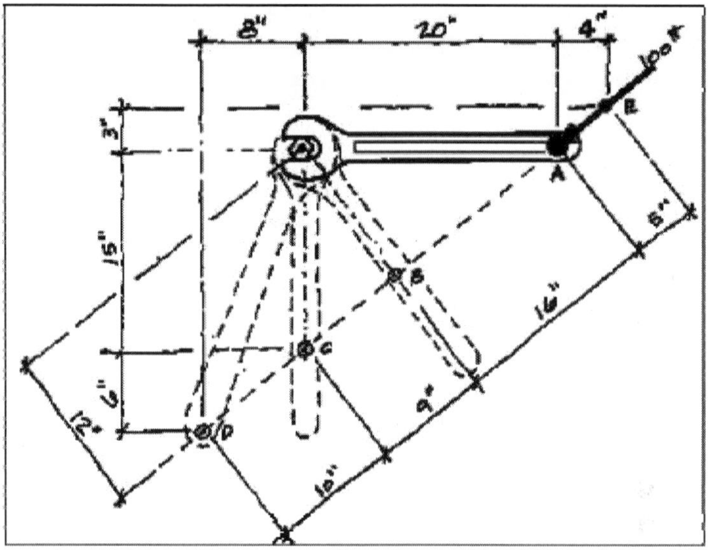

First Consider the 100 Pound Force

Since the line of action of the force is not perpendicular to the wrench at A, the force is broken down into its orthagonal components by inspection. The line of action of the the 100 pound force can be inspected to determine if there are any convenient geometries to aid in the decomposition of the 100 pound force.

The 4 inch horizontal and the 5 inch diagonal measurement near point A should be recognised as belonging to a 3-4-5 triangle. Therefore, Fx = -4/5(100 pounds) or -80 pounds and Fy = -3/5(100 pounds) or - 60 pounds.

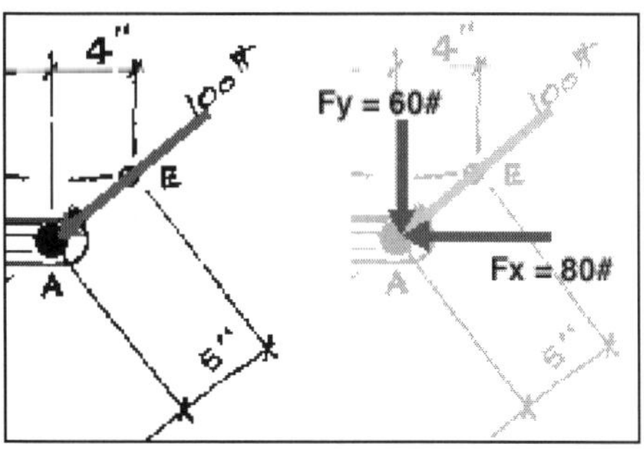

Consider Point A

The line of action of Fx at A passes through the handle of the wrench to the bolt (which is also the centre of moments). This means that the magnitude of the moment arm is zero and therefore the moment due to FAx is zero. FAy at A has a moment arm of twenty inches and will tend to cause a positive moment.

FAy d = (60 pounds) (20in) = 1200 pound-inches or 100 pound-feet

The total moment caused by the 100 pound force F at point A is 1200 pound-inches.

Consider Point B

At this point the 100 pound force is perpendicular to the wrench. Thus, the total moment due to the force can easily be found without breaking it into components.

FB d = (100 pounds) (12in) = 1200 pound-inches

The total moment caused by the 100 pound force F at point B is again 1200 pound-inches.

Consider Point C

The force must once again be decomposed into components. This time the vertical component passes through the centre of moments. The horizontal component FCx causes the entire moment.

FCx d = (80 pounds) (15 inches) = 1200 pound-inches.

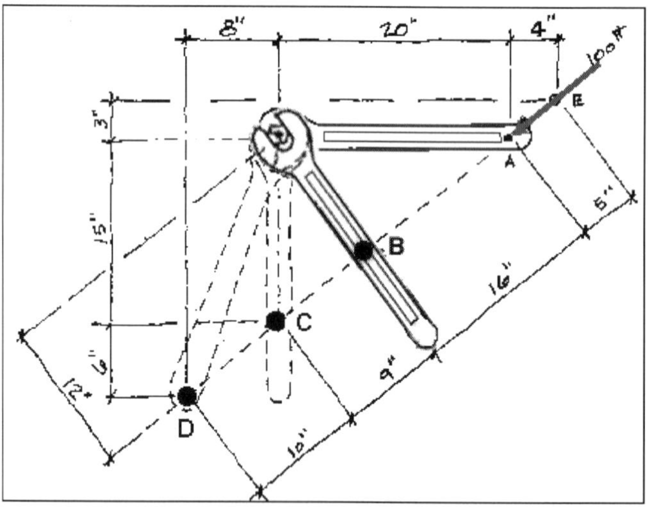

Consider Point D

The force must once again be decomposed into components. Both components will contribute to the total moment.

FDx d = (80 pounds) (21inches) = 1680 pound-inches

FDy d = (60 pounds) (8 in) = -480 pound-inches

Note that the y component in this case would create a countre clockwise or negative rotation. The total moment at D due to the 100 pound force is determined by adding the two component moments. Not surprisingly, this yields 1200 pound-inches.

Consider Point E

Following the same procedure as at point D.

FEx d = (80 pounds) (3 in) = -240 pound-inches

FEy d = (60 pounds) (24 in) = 1440 pound-inches

However, this time Fx tends to cause a negative moment. Once again the total moment is 1200 pound-inches.

At each point, A, B, C, D and E the total moment around the bolt caused by the 100 pound force equalled 1200 pound-inches. In fact, the total moment would equal 1200 pound-inches at ANY point along the line of action of the force. This is Varignon's Theorem.

Non-Concurrent, Non-Parallel Force Systems

The principles of equilibrium are also used to determine the resultant of non-parallel, non-concurrent systems of forces. Simply put, all of the lines of action of the forces in this system do not meet at one point. The parallel force system was a special case of this type. Since all of these forces are not entirely parallel, the position of the resultant can be established using the graphical or algebraic methods of resolving coplanar forces discussed earlier or the link polygon.

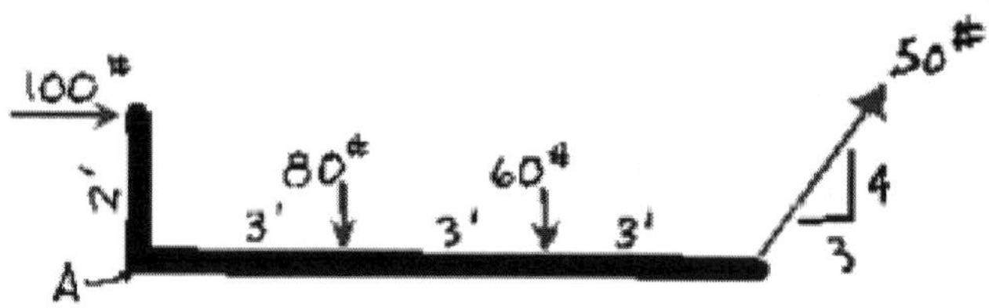

There are a number of ways in which one could resolve the force system that is shown. One graphical method would be to resolve a pair of forces using the parallelogram or triangle method into a resultant. The resultant would then be combined with one of the remaining forces and a new resultant determined, and so on until all of the forces had been accounted for. This could prove to be very cumbersome if there is a great number of forces. The algebraic solution to this system would potentially be simpler if the forces that are applied to the system are easy to break into components. The algebraic resolution of this force system is illustrated in the example problem.

The Link Polygon is a graphical method to determine the resultant of non-concurrent, non-parallel force systems. It depends on an accurate graphical representation of all of the forces acting on a system. Essentially, each of the given forces is successively replaced by two components that are arranged so that one component of the force is equal in value to a component of the succeeding force. When all of the components have been combined in this way, the remaining force is the resultant of the system. This is accomplished by having a common "polar point" within the force polygon from which the components of the forces are generated. This point can be located anywhere in space, but is often chosen so that the link polygon will fit on the free body diagram.

Couples

A special case of moments is a couple. A couple consists of two parallel forces that are equal in magnitude, opposite in sense and do not share a line of action. It does not produce any translation, only rotation. The resultant force of a couple is zero. BUT, the resultant of a couple is not zero; it is a pure moment.

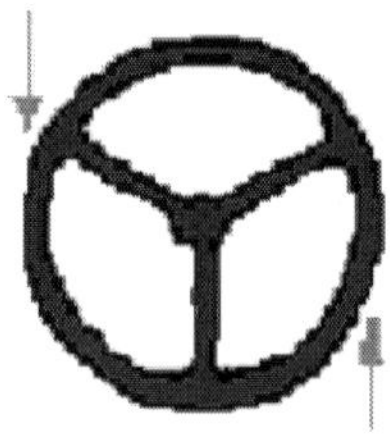

For example, the forces that two hands apply to turn a steering wheel are often (or should be) a couple. Each hand grips the wheel at points on opposite sides of the shaft. When they apply a force that is equal in magnitude yet opposite in direction the wheel rotates. If both hands applied a force in the same direction, the sum of the moments created by each force would equal zero and the wheel would not rotate. Instead of rotating around the shaft, the shaft would be loaded with a force tending to cause a translation with a magnitude of twice F. If the forces applied by the two hands were unequal, there would again be an unbalanced force creating a translation of the "system." A pure couple always consists of two forces equal in magnitude.

The moment of a couple is the product of the magnitude of one of the forces and the perpendicular distance between their lines of action. M = F x d. It has the units of kip-feet, pound-inches, KN-metre, etc. The magnitude of the moment of a couple is the same for all points in the plane of the couple. A couple may be moved anywhere in its plane or a parallel plane without changing its external effect. The magnitude of the couple is independent of the reference point and its tendency to create a rotation will remain constant. This can be illustrated with the simple illustration of a bar with a length d that is pinned at its midpoint. Two parallel forces of equal magnitude, opposite in sense are applied at the ends of the bar. The magnitude of the moment generated by the couple of the forces F, relative to the pin in the illustration, is equal to

$$(F)(d/2) + (F)(d/2) = (F)(d)$$

The magnitude of the couple of the forces F relative to point "O" is

$$(F)(d+x) - (F)(x)$$

$$(F)(d) + (F)(x) - (F)(x) = (F)(d)$$

Again, it can be seen that the magnitude of the couple is independent of the reference location. It is always equal to (F)(d)!

The resultant of a number of couples is their algebraic sum. A couple CANNOT be put in equilibrium by a single force! A couple can only be put in equilibrium by a moment or another couple of equal magnitude and opposite direction anywhere in the same plane or in a parallel plane. If a single force is added to the system that balances the sum of the moments, one of the other two equations of equilibrium will not be satisfied. A couple maintains the internal equilibrium of a simple beam. The concept is very important to the further study of structural behaviour.

Supports

Structural elements carry their loading to other elements or the ground through connections or supports. In order to be able to analyse a structure it is necessary to be clear about the forces that can be resisted at each support. The actual behaviour of a support or connection can be quite complicated. So much so, that if all of the various conditions were considered, the design of each support would be a terribly lengthy process. And yet, the conditions of the supports is very important to the behaviour of the elements which are being supported.

In order to facilitate the analysis of a structure, it is often necessary to idealize the behaviour of a support. This is similar to the massless, frictionless pulley in physics homework problems. Even though these pulleys do not exist, they are useful to enable learning about certain issues. It is important to realize that all of the graphical representations of supports are idealisations of a real connection. Effort should be made to search out and compare the reality with the graphical and/ or numerical model. It is often very easy to forget that the reality can be strikingly different than what is assumed in the idealisation!

The four types of supports that can be found in structures are; roller, frictionless surface, pinned, and fixed. The type of support affects the forces and moments that are used to represent these supports. It is expected that these representative forces and moments, if properly calculated, will bring about equilibrium in the structural element. There is not a single accepted graphical method to represent each of these support types. However, no matter what the representation looks like, the forces that the type can resist is indeed standardised. Almost all supports can be assigned to one of the four types.

Roller Supports

Roller supports are free to rotate and translate along the surface upon which the roller rests. The surface can be horizontal, vertical, or sloped at any angle. The resulting reaction force is always a single force that is perpendicular to, and away from, the surface. Roller supports are always found at at least one end of long bridges so that forces due to thermal expansion and contraction are minimalised. These supports can also take the form of rubber bearings which are designed to allow a limited amount of lateral movement. A roller support cannot provide any resistance to lateral forces. The representation of a roller support includes one force perpendicular to the surface.

Frictionless Supports

Frictionless surface supports are similar to roller supports. The resulting reaction force is always a single force that is perpendicular to, and away from, the surface. They too are often found as supports for long bridges or roof spans. These are often found supporting large structures in zones of frequent seismic activity. The representation of a frictionless support includes one force perpendicular to the surface.

Pinned Supports

A pinned support can resist both vertical and horizontal forces but not a moment. They will allow the structural member to rotate, but not to translate in any direction. Many connections are assumed to be pinned connections even though they might resist a small amount of moment in reality. It is also true that a pinned connection could allow rotation in only one direction; providing resistance to rotation in any other direction. The knee can be idealised as a connection which allows rotation in only one direction and provides resistance to lateral movement. The design of a pinned connection is a good example of the idealisation of the reality. A single pinned connection is usually not sufficient to make a structure stable. Another support must be provided at some point to prevent any rotation of the structure. The representation of a pinned support includes both horizontal and vertical forces.

Fixed Supports

Fixed supports can resist vertical and horizontal forces as well as a moment. Since they restrain both rotation and translation, they are also known as rigid supports. This means that a structure only needs one fixed support in order to be stable. All three equations of equilibrium can be satisfied. A flagpole set into a concrete base is a good example of this kind of support. The representation of fixed supports always includes two forces (horizontal and vertical) and a moment.

Free Body Diagrams

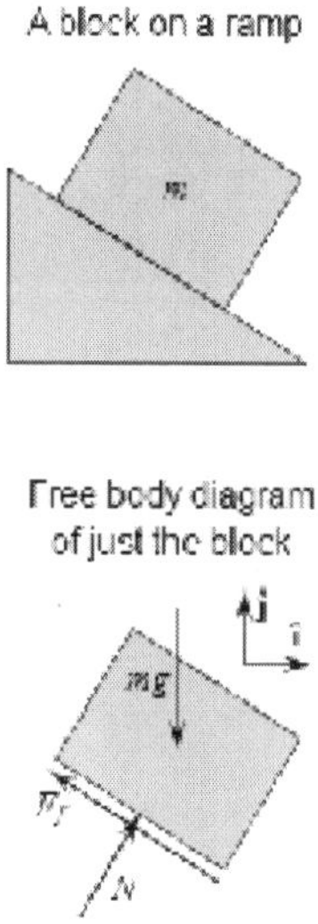

***Figure:** Block on a ramp (top) and corresponding free body diagram of just the block (bottom). For equilibrium, the line of action of the three force arrows must intersect at a common point.*

A free body diagram, sometimes called a force diagram, is a pictorial device, often a rough working sketch, used by engineers and physicists to analyze the forces and moments acting on a body. The body itself may consist of multiple components, an automobile for example, or just a part of a component, a short section of a beam for example, anything in fact that may be considered to act as a single body, if only for a moment. A whole series of such diagrams may be necessary to analyze forces in a complex problem. The free body in a free body diagram is not free of constraints, it is just that the constraints have been replaced by arrows representing the forces and moments they generate.

Purpose

Drawing a free body diagram can help determine the unknown forces on, moments applied to, and equations of motion of, the body and thus help to analyse a problem in statics or dynamics. In analysis of structures, free body diagrams for a component of a structure or, part thereof, are used in determining shear forces and bending moments.

Construction

A free body diagram, an FBD, is not meant to be a scaled drawing. Rather it is a working sketch open to modification as one works through the problem and typically one needs to have seen through the problem before one arrives at a satisfactory diagram. There is an element of art, an inherent flexibility in the whole process. There is no hard and fast algorithm. The iconography of a free body diagram - not only how it is drawn but also how it is interpreted - depends crucially on how a body is modelled.

Modelling the Body

A body may be modelled in three ways:

(i) *A particle:* This model may be used when any turning effects are zero or have zero interest even though the body itself may be extended. The body may be represented by a small symbolic blob and the diagram reduces to a set of concurrent arrows. A force on a particle is a bound vector.

(ii) *Rigid Extended:* Stresses and strains are of no interest but turning effects are. A force arrow should lie along the line of force, but where along the line is irrelevant. A force on an extended rigid body is a sliding vector.

(iii) *Non-rigid Extended:* The point of application of a force becomes crucial and has to be indicated on the diagram. A force on a non-rigid body is a bound vector. Some engineers use the tail of the arrow to indicate the point of application. Others use the tip.

Example: A Body in Free Fall

The foregoing remarks can be elucidated with the help of an example. Consider a body in free fall in a uniform gravitational field. The body may be

(i) a particle. It is enough to show a single vertically downward pointing arrow attached to a blob.

(ii) rigid extended. A single arrow suffices to represent the weight W even though gravitational attraction acts on every particle of the body. The arrow must lie along a line through the centre of gravity. Where exactly along the line the arrow should lie is a matter of convenience. A popular choice is to locate the arrow at the centre of gravity but strictly speaking there is no need to do so in rigid body analysis.

Example: A Block on an Inclined Plane

A simple free body diagram, shown above, of a block on a ramp illustrates this.

All external supports and structures have been replaced by the forces they generate. These include:

mg: the product of the mass of the block and the constant of gravitation acceleration: its weight.

N: the normal force of the ramp.

Ff: the friction force of the ramp.

The force vectors show direction and point of application and are labeled with their magnitude.

It contains a coordinate system that can be used when describing the vectors.

Some care is needed in interpreting the diagram. The line of action of the normal force has been shown to be at the midpoint of the base but its true location can only be found if sufficient further data is given. The diagram as it stands would need to be modified were we told that the block is in equilibrium.

There is a potential difficulty also with the arrow representing friction. The engineer who drew this diagram has used the tip of the arrow to indicate the point of application of a force. Now, the tip of the friction arrow is at the highest point of the base. The intention however is not to indicate that the friction acts at that point. The engineer in this instance has assumed a rigid body scenario and that the friction force is a sliding vector and thus the point of application is not relevant.

The engineer has tried to indicate that the friction acts all along the whole base by drawing an arrow all along the base but such artistic ploys are a matter of personal choice.

Chapter 5

Design of Abutments

They are the end supports of the superstructure, retaining earth on their back. They are built either with masonry stone or brick work or ordinary mass concrete or reinforced concert. The top surface of the abutment is made flat when the superstructure is of trusses or girders or semi circular arch. In case of segmental or elliptical arch type of superstructure, the abutment top is made skew. Weepholes are provided at different levels through the body of the abutment to drain off the retained earth.

The salient features of bridge abutments are listed below :

a. Height : The height of the abutment is kept equal to that of the piers.

b. Abutment Batter : The water face of the abutment is usually kept vertical or could be given a batter of 1 in 12 1 in 24 as for piers. The face retaining earth is given a batter of 1 in 6 or may be stepped down.

c. Abutment Width : The top width of the abutment should provide enough space for the bridge seat and for the construction of a dwarf wall to retain earth up to the approach level.

d. Length of Abutment : The length of abutment is kept at least equal to the width of the bridge.

e. Abutment Cap : The design is similar to that of pier cap.

Abutments can be spill through or closed. The spillthrough abutment generally has a substantial berm to help restrain embankment settlement at the approach of the structure. Approach embankment settlement can also be accommodated by approach slabs to eliminate bumps at the bridge ends. Closed abutments partially or completely retain the approach embankments from the spilling under the span. Bridges or several spans require expansion at the abutments. Therefore, they are not usually required to resist the longitudinal forces that develop.

In general, the loads applied to abutments are the vertical reactions, the lateral earth pressure and an accompanying surcharge effect. Sometimes the friction or shear force produced by expansion bearings is significant and worthy of consideration and can have an additive effect on the earth pressure and surface forces.

Broadly abutments are classified under the following two categories:

1. Abutments with wing walls
2. Abutments without wing walls.

Abutments with Wing Walls

In this category, the abutments are provided with wing walls which may be :

a. Straight wing walls or
b. Splayed wing walls, or
c. Return wing walls.

The details of a standard gravity abutment with wing walls. In this type of abutment the wing walls can be either straight or splayed. The abutments wing splayed; with walls are used across a river to provide smooth entry and exist for the water.

The wing walls not only withstand the earth pressure but also the impact of live load over them. The abutments with return wing walls are known as U abutments.

In this type the abutment is extended at right angles on both the ends to some distance to protect earth work. These are used where banks are steep and rocky.

The abutments with wing walls suffer from the following disadvantages :

a. They tend to restrict the flood prism and hence increase score and raise upstream flood level.
b. The scour may require deeper foundations
c. Foundations may have to be built-in the wet. These types are indicated for small heights. Special attention must be paid to the junction of the wing and abutments walls to prevent cracks.

Abutments without Wing Walls

In this category, the following types of abutments are in general use:

a. Buried abutments
b. Box abutments
c. Tee abutments
d. Arch abutments.

Buried Abutments : This type of abutment is generally built prior to the placing of the fill. Since it is filled on both sides the earth pressure is low. Superstructure erection can begin before placement of fill.

Box Abutment : This employs a short span of bridge built integral with columns to act as a frame and resist earth pressure of the approaches. It is most often used for overpass work where the short span may be employed for pedestrian passage.

Tee Abutments : This type looks like T in plan and has now become obsolete see.

Arch Abutments : This type of abutment is used where arches are employed because of their economy in certain conditions. The high inclined skewback thrusts are difficult to handle unless the abutment can be seated in rock. Therefore, they are often used for span over gorges.

Design of Abutments

These are designed in a similar manner as piers. Except for the fact that they act as a retaining structure and are subjected to an additional force i.e. earth pressure. The magnitude of the earth pressure depends upon the type and moisture content of the backfill material. Therefore, it is most difficult to assess correctly the magnitude of the earth pressure. Hence the earth filling behind the abutment merits special attention. Normally, a porous backfill of about 60 cm thickness should be provided immediately behind the abutment with 15 cm diametre weepholes placed at suitable intervals with a gentle slope so as to be 15 cm above normal water level on the vent side to ensure drainage for the filling material. Impact effect from moving loads is not usually accounted for since any vibration is supposed to be dissipated through embankment.

The conditions of stability for abutments are similar to the one as described for the piers. Cares should be taken during abutment design to check the stability of the abutment structure when loaded with full earth pressure plus live load surcharge without superstructure dead load reaction.

If an unreasonably low factor of safety against overturning is attained, backfilling should be limited until the superstructure dead load is in place. In some abutment designs it is advisable to check heel pressure with a maximum reaction and one half the normal earth pressure. An experienced designer can look over his design problem and determine whether this additional check is necessary.

Determination of Flood Discharge

The most important factor to be decided in the design of bridges is the determination of the waterway required for passing the normally expected highest flood discharge. There are a number of factors which affect the runoff and thus the flood discharge from a catchment area served by a bridge or a culvert. They can be broadly divided into three viz.

- rainfall and weather characteristics;
- terrain characteristics; and
- stream characteristics.

A further step is to apply the method of statistics and probability. This applicable for a catchment for which data of the highest annual floods are available over a period of at least 25 to 30 years. A probability of once in 50 years up to once in 100 years is used depending upon the catchment.

While fixing the design flood discharge, the recommended method is to use at least two of the methods and arrive at a figure which is the maximum of the two or 1.5 times the minimum whichever is less. Though a certain amount of risk can be taken by providing a waterway a little less than what is required for the worst expected flood, such a risk cannot be taken for fixing the foundation for the structure.

One of the essential input data for the bridge design is a fair assessment of the maximum flow which could be expected to occur at the bridge site during the design life period of the bridge. This faulty determination of flood discharge has led to failure of many hydraulic structures. The India Roads Congress has recommended that the maximum discharge which a bridge on a natural stream should be designed to pass should be determined by a consideration of at least two of the following methods:

- From the rainfall and other characteristics of the catchment area.
- By use of an empirical formula applied to that region, or
- By a rational method, provided it is possible to evaluate for the region concerned the various factors employed in the method
- From the hydraulic characteristics of the stream such as cross sectional area, and slope of the stream allowing for velocity of flow.
- From the records available, if any, of discharges observed on the stream at the site of the bridge, or at any other site in its vicinity.

Vertical Clearance above HFL

For high level bridges, a vertical clearance should be allowed between the highest flood (HFL) and the lowest point of the superstructure. This is required to allow for any possible error in the estimation of the HFL and the design discharge. It also allows floating debris to pass under the bridge without damaging the structure. It is, therefore, logical to provide higher values of clearance for greater discharge as in table given below, based on the provisions of IRC:

Discharge (cumecs)	*Min. vertical clearance (mm)*
upto 0.3	150
0.3 - 3	450
3 - 30	600
31 - 300	900
301 - 3000	1200
ovcr 300	1500

For arched bridges, the clearance below the crown of the intrados of the arch should not be less than one tenth of the maximum depth of the water plus one third of the rise of the arch intrados. For structures provided with metallic bearings, the clearance between the base of the bearings and the highest flood level taking afflux into account is not to be less than 500 mm. For irrigation channels, the vertical clearance may be relaxed at the discretion of the engineer-in-charge.

The difference between vertical clearance and freeboard is sometimes not clearly understood. While vertical clearance is the difference in the level between HFL and the lowest point of the superstructure, freeboard is associated with the approaches and guide bounds. The freeboard at any point is the difference between the highest flood level after allowing for afflux, if any, and the formation level of road embankment of the approaches or top level of guide bounds at that point. For high level bridges, the freeboard should not be less than 600 mm.

Scour Depth

Scour of stream bed occurs during the passage of a flood discharge, when the velocity of the stream exceeds the limiting velocity that can be withstood by the particles of the bed materials. The scour is aggravated at the nose of the piers and at bends.

The maximum depth of scour should be measured with reference to existing structures near the proposed bridge site, if this is possible. Such soundings are best done during or immediately after a flood. Due allowance should be made in the observed values for additional scour that may occur due to the designed discharge being greater than the flood discharge for which the scour was observed, and also due to increased velocity due to obstruction to flow caused by the construction of the bridge. The parametre of scour depth plays a very important role in the design of foundations of river bridges, specially those resting on soil strata.

Where the above practical method is not possible, the mean depth of scour may be computed by the following equation for natural streams in alluvial beds.

$$d_{sm} = 1.34\ (D^2_b\ /\ K_{sf})^{0.33}$$

where d_{sm} = the mean depth of scour below HFL in metres

d_b = discharge in m³/s per m width obtained as the total design discharge divided by the effective linear waterway

K_{af} = silt factor for a representative sample of the bed material, taken as 1.76 times the square root of the particle size in mm.

AFFLUX

Another term one has to be familiar with while designing the waterway through bridge is 'afflux'. It can be defined as a rise of heading up of water level (above normal) on the upstream side of a bridge or obstruction. It is caused when the effective linear waterway at the obstruction is less than the natural width of the stream immediately on the upstream side of the obstruction. In case of a bridge where there is no reduction in the overall width of the waterway over the natural width of the stream, such an afflux can also be caused by the obstruction produced by piers and projecting abutments. The greater the afflux, the more will be the velocity produced through the obstruction.

An estimation of the afflux is necessary to determine the bottom line of the bridge deck after providing for adequate clearance under the bridge and for fixing levels of the approach road/track and protection works after allowing for adequate free board as well as for determining the velocity for the purpose of designing of the foundation and any bed protection works.

Afflux is given by the formula.

$h = (V^2\ W^{2)} / 2g\ c^2L^2 - 1$

where h = afflux

v = velocity of normal flow in the stream

g = acceleration due to gravity

W = width of stream at HFL (regime width)

L = Linear waterway under the bridge (all expressed in fps units)

c = coefficient of discharge through the bridge, taken as 0.7 for sharp entry and 0.9 for bell mouthed entry.

The afflux should be kept minimum and limited as far as possible to 150 mm. The safe velocities for different types of soil under the bridge are given below :

Loose clay or fine sand upto 0.5 m/s

Coarse sand 0.5 to 1.0 m/s

Find gravel, sandy or stiff clay 1.0 to 1.5 m/s

Coarse gravel, rocky soil 1.5 to 2.5 m/s

Boulders, rock 2.5 to 5.0 m/s

In case the velocity exceeds those specified above, a suitable flooring with proper drop wall should be provided under the bridge as indicated in Fig. 5.2

The depth of drop wall should be such that does not get undermined. If a flooring would not have been provided, depth of scour downstream would be

$D_s = 0.30h^{0.2}\ q^{0.57} / d_m$

where h is the afflux

q is the discharge/unit width in cusecs

d_m is the effective diametre of soil in millimetres

Drop wall should be made sufficiently deeper than this.

Case Study

The following case study for approach to hydrological studies concerning bridges has been taken from an upgradation project for converting two lanes to four lanes along a section of National Highway No. 8, India in the states of Gujarat and Maharashtra.

Approach used to Carry out Assessment of Design Discharge at Bridge Sites

As per the Khosla Committee of Engineers recommendation a design flood of 50 years return period has been taken for fixing the waterway of bridges.

Indian Road Congress (IRC) also lays down similar recommendations. IRC also recommends that the design discharge of the foundation should be a given percent more than the waterway design discharge depending upon the catchment area. As per the code the design discharge will be increased as per following:

The design discharge in this project has been estimated by application of hydro meteorological approach as well as by using empirical relation as is in practice in that region. In addition discharge is estimated at observed HFL / design HFL (high flood level) by Area Velocity method.

In addition for sake of comparison, discharge as observed by the Irrigation Department on the rivers crossing on the National Highway has also been taken into account. Besides design discharge for waterway of some bridges as given in PWD records has also been mentioned.

Hydro Meteorological Approach

The existing two lane bridges are in existence for the last 40 - 45 years. However, there is no record of annual HFL / discharge at the site. In such a situation, the 50 year return period flood value has been estimated as per guidelines of Central Water Commission of the West Coast region.

The method given in the CWC report is based on hydro - meteorological approach and consists of development of synthetic unit hydrography with catchment parametres and application of design rainfall to unit hydrogaphy. Study on design rainfall is included in the aforesaid report.

Empirical Formula Approach

In Bombay Presidency, Inglis formula (IRC-SP-13) has been used for assessment of discharge of the rivers. The formula is as below :

$$Q = \frac{125\,M}{\sqrt{M} + 10}$$

Where, Q = maximum flood discharge in cumecs

M = the area of catchment in Sq.Km.

Discharge Estimation and Results

Discharge Estimation for the Bridges Having Catchment Area more than 25 Sq.km (U/G Method)

This is based on development of synthetic unit hydrographic using basin characteristics parametres, and application of design storm to get the design flood

1 hour unit hydrography parametres :

U.G. = Unit Graph

t_r = Unit Rainfall duration adopted in a specific study (hr.)

T_m = Time from the start of the rise to the peak of the U.G. (hr.)

Q_p = Peak Discharge of Unit Hydrography (cumec)

t_p = Time from the Centre of Effective Rainfall duration to the U.G. Peak (hr.)

W_{50} = Width of the UG measured at 50% of peak discharge ordinate (hr.)

W_{75} = Width of the rising limb of UG measured at 50% of peak discharge ordinate (hr.)

W_{R75} = Width of the rising limb of UG measured at 75% of peak discharge ordinate (hr.)

T_B = Base width of Unit Hydrography (hr.)

A = Catchment Area (Sq.km)

q_p = Qp/A = Cumec per Sq.km

Discharge Estimation for the Catchment Area less than 25 Sq.km (Rational Method) [Ref IRC-SP-13]

The estimation is based on rational formula which is as below :

Design discharge $Q = AI_0\grave{U}$

Where A = catchment area in hectares

I_0 = *One hour rainfall intensity in cm/hr*

$$I_0 = \frac{F(1+1/T)}{2}$$

F = The 24 hour storm rainfall in cm corresponding to 50 year frequency

T = duration of storm in hour (24 hour)

$$\wedge = \frac{0.056\ x\ fP}{(t_c + 1)}$$

P = Percentage coefficient of runoff for the catchment characteristics

f = Fraction of maximum point intensity

$t_c = \{0.87\ L^3\}^{0.385} / H$

t_c = *the concentration time in hour*

L = the distance from the farthest point from periphery of the catchment (critical point) to the culvert in km

H = the fall in level from the critical point to the culvert in metre

P = 0.70 for plateaus, lightly covered catchments

= 0.80 for rock, steep but wooded catchments

Discharge Estimation by Empirical Formula

In Bombay Presidency, Design discharge

$$Q = \frac{125M}{\sqrt{}}$$

M is the catchment area in sq.km

Area Velocity Method

The discharge is calculated as per Manning's formula :

$V = 1/n\ R^{2/3}\ S^{1/2}$

Where V = Velocity of river/stream

n = rugosity coefficient

R = Hydraulic mean depth

= Cross Sectional Area = A/P

Wetted perimetre

S = Longitudinal bed slope

=> *Q = AXV*

Q = Discharge

Value of n has been estimated as per IRC:AP:1990,

Slope has been estimated from the topographical sheet

Cross-sectional area has been estimated at observed HFL as well as at PWD designed HFL.

Observed Discharge at Irrigation G & D Sites

The Irrigation Department of Gujarat Government is maintaining gauge and discharge sites on the some of the rivers upstream of their National Highway crossing points. Design discharges of some bridges have been mentioned in PWD records.

Recommended Design Discharge

The recommended has been based on following conditions :

- Design discharge as given by PWD has been adopted when the discharge estimated by the other methods are close to the discharge given by PWD.
- In cases when the discharge given by other methods are higher than the PWD design discharge, weighted has been given to discharge estimated by hydrometrological approach in case of bridge site having catchment area more than 25 sq.km.
- In case of catchment area less than 25 sq.km where design discharge has been dictated by PWD and close to Rational formula of Manning's formula, the discharge given by PWD has been recommended.
- In case where the estimate given by Rational or Area Velocity Method is significantly higher than the PWD design discharge, weighted has been given to Rational formula or Area-Velocity Method, whichever is closer to PWD design. It is significantly lower than the PWD discharge, the design discharge has been taken as 1.5 times of the second highest value of discharge.
- In case where design discharge has not been indicated by PWD, the design discharge as estimated by Area-Velocity or Rational or Inglis have been compared and highest of these three estimates has been adopted. It the difference between the highest and second highest is very large, the design discharge has been taken as 1.5 times of second highest value of discharge.
- In some cases, PWD design discharge and discharge estimated by Inglis formula are on very much higher side as compared to discharges estimated by other methods. In such cases the

discharge estimated by U G Method of Rational Method and Area-velocity method have been compared and accordingly recommended.

Sub-Surface Investigation

The determination of a reasonably accurate soil profile at each of the proposed bridge sites is essential for correctly deciding the location and type of foundation. The subsoil investigation should aim to provide adequate data to the designer at a cost consistent with the value of the information. Availability of correct and reliable data would enable the construction contractors to estimate their costs correctly and to plan their work intelligently, resulting in a better job at a lower cost. Defects in bridge structures attributable to serious errors in subsoil exploration cannot be easily rectified later. Guidelines for subsoil exploration are available in IRC Bridge Code - Section VII.

It has been observed that most of the failures of structures are as a consequence of the failure of foundations. As such an accurate assessment of the physical properties of the founding strata at the proposed bridge site is necessitated in order to decide upon the location and type of foundation with a certain degree of correctness. The field and laboratory investigations required to obtain the necessary soil data for the design are called soil exploration. Due to the experience gained and to the better knowledge in design of foundations the number of such failures have shown a fall in recent years. To arrive at an economical and safe design, quantitative evaluation of soil properties needs to be carried out properly. Varying nature of soils at different sites or even that same site results in making the problem of soil exploration more complex. The extent of investigation for each job will depend upon the information already available regarding the soil in the area and on the relative importance of the bridge structure.

Sub-Surface Investigation

The principal requirements of a complete investigation can be summarised as follows :

1. Nature of the soil deposits up to the sufficient depth.
2. Depth thickness and composition of each soil stratum.
3. The location of ground water.
4. Depth of rock and the composition of rock strata.
5. The engineering properties for the soil and the rock strata that affect the design of the structure.

To establish the extent of distribution of different soils both in the horizontal and vertical directions during soil exploration the following methods may be used :

1. By use of open pits
2. By making bore holes and taking out samples
3. By soundings
4. By use of geophysical methods
5. Use of open pits is expensive. Information of specialised nature can be obtained by this method because it gives a complete picture of the whole area. Bore holds are drilled by Auger borings, displacement borings, wash borings, percussion drilling and continuous sampling method. The sounding rod method of suitable only as a preliminary investigation to determine the location of rock at shallow depths.

The Sub Surface Investigations for Bridges can be Carried out in Two Stages, Preliminary and Detailed.

The aim of the preliminary investigations to get a general idea about the nature of soil strata. The methods used may include the study of existing geological information, besides field investigations with sounding rods, auger borings, wash borings, test pits and possibly geophysical methods.

Sounding rods consist of solid bars of tool steel of 15 to 20 mm diametre. The bottom end of the first length (usually about 3 m) is pointed. The top end is threaded so that additional lengths of about 2 m can be coupled. The sounding rod is first churned into the soil by hand to about 2 m and then driven with a 50 km. hammer. Additional sections are added as necessary. The resistance to driving is to be carefully interpreted to give a rough idea of the type of soil met with. This method is suitable only as a preliminary investigation to determine the location of rock at shallow depths.

Earth auger and post hole auger are used singly for shallow depths and in conduction with a 60 mm pipe casing and additional coupled rods for larger depths upto about 15 to 20 m. Samples are taken out in the augers and examined. When samples will not stick to the auger, wash borings are used. Here, a water jet is forced down a wash pipe (drill rod itself, if hollow). The water rises through the annular space between the casing and the wash pipe, carrying fragments of soil which is collected and examined.

After a study of the data collected in the preliminary investigation, the bridge site, the type of structure with span layout, and the location and type of foundations are tentatively determined. The detailed exploration should cover the entire length of the bridge and should extend at either end of the bridge for a distance of about twice the depth of foundation of the abutment. The depth of investigation should generally extend to a depth about twice the width of foundation below the level of the foundation below the level of foundation, and should preferably, be extended into firm soil or rock.

Borings have to be taken at suitable intervals, including the probable locations of abutments and piers. The data required are (i) nature of soil deposit, (ii) depths and thickness of soil strata, (iii) location of ground water table, (iv) depth to rock bed, and (v) engineering properties of soils and rock. Standard penetration test (SPT) values expressed as number of blows per 300 mm penetration are also determined at different depths below the surface. Core drills are used to take out samples of rock to examine the thickness of rock layer, and to ensure that the layer reached is not a boulder but rock of adequate thickness.

Boring charts for the boreholes are first drawn individually listing the various details. By plotting the data for a number of boreholes along the cross section of a stream and connecting the corresponding points for each layer, the subsoil profile for a proposed site can be obtained. The proposed foundation level is indicated by a dashed line. If the locations of abutments and piers are known, and if the foundations are rested at different levels, such levels may be indicated at the abutment and pier locations instead of one level.

If the approaches involve high embankments, boring should also be taken along the approaches. The importance of this requirement is not realised adequately. In a particular major bridge of cantilever type, high embankments of about 12 m were proposed instead of viaduct spans with a view to reduce the overall cost. The presence of clayey soil below the embankments was not realised (for want of borings along the approaches) until the failure of the embankment. The subsequent repairs involved introduction of viaduct spans of different structural layout, resulting in permanent damage to the aesthetics at both ends of the bridge.

Geophysical methods are used for quick preliminary exploration over large areas. This gives reliable information about the depth, thickness and composition of rock.

A general idea of the type of strata is obtained through preliminary trials bores. The standard penetration tests conducted at the proposed bridge site which gave an idea of the resistance of soil to penetration and to detect loose pockets, if any. With the advent of large span, continuous and rigid frame structures, the aspects of settlement of piers may alter the magnitude and even the nature of stresses in the structure considerably.

No hard and fast rule can be laid down for the extent of exploration required. An important aspect of exploration is the depth and spacing of the borings. Initially the borings may be spaced far apart and they should be taken to depths considered adequate to disclose all strata which could consolidate or fail by shear under the loads imposed by the structure. This depth may be about one to two times and anticipated width of the foundation below the anticipated depth of foundation. The spacing can be closer or wider depending upon the heterogeneity or homogeneity of the deposit indicated by the initial borings.

Forces to be Accounted for in the Analysis for the Design of a Bridge

Main Forces

The following is a list of the main forces whose effects should be analysed to estimate the load-effects (moments shears, etc.) at all critical sections in the structure. Only then the structure should be design for these load-effects to decide the section size, reinforcements, prestress, etc., so as to resist these forces at the specified stress levels, and serviceability criteria (crack-widths, deflections, etc.) :

I. Dead load of the structure (in some cases self-weight may come in stages)

II. Live load (on roadway, cycle tracks and footpaths). Ref. Clause 207, 208 and 210 of IRC:6-1966

III. Impact effect of moving live load. Ref. Clause 211 of IRC:6-1966.

IV. Braking force or Longitudinal forces (generated by the application of brakes on the live load). Ref. Clause 214 of IRC:6-1966.

V. Wind load Ref. Clause 212 of IRC:6-1966.

VI. Seismic force Ref. Clause 222 of IRC:6-1966.

VII. Lateral horizontal loads on parapets and kerbs. Ref. Clause 209 of IRC:6-1966.

VIII. Centrifugal force in horizontally curved decks. Ref. Clause 215 of IRC:6-1966.

IX. Flood water current forces in the bridge longitudinal and transverse directions. Ref. Clause 213 of IRC:6-1966.

X. Effect of afflux head (created locally on a pier as it obstructs the flood flow).

XI. Effect of crosscurrent force in bridge longitudinal direction (the effect of about 25 cm static head of water across the pier, looking at the bridge length), where the river bed is unerodible (rock) at scour level. (This force is an alternative to the flood water force in bridge longitudinal direction.)

XII. Buoyancy effect (floatation). Ref. Clause 216 of IRC:6-1966,

XIII. Earth pressure Ref. Clause 217 of IRC:6-1966.

XIV. Self-induced horizontal force caused at bearings by movement/ rotation of deck due to temperature variation, creep and shrinkage of deck concrete elastic-shortening of deck due to prestress etc.) Ref. Clause 218 of IRC:6-1966.

XV. Thermal effect (comprising (i) the effect of non-linear distribution of temperature through the deck depth (leading to eigen stress resulting from the difference between final linear thermal strain gradient and the unrestrained non-linear thermal strain gradient), and (ii) the indeterminacy effect (comprising (a) effect of restraint to change in body mean temperature and (b) effect of restraint to angular movements at the supports)). Ref. Clause 218 of IRC:6-1966.

XVI. Secondary effects (e.g. effect of eccentric connections, and shrinkage and creep of deck concrete). Effect of possible differential settlement of supports. Ref. Clause 220 of IRC:6-1966.

XVII. Loads resulting from temporary erection conditions and partial span-dislodgment conditions. Ref. Clause 221 of IRC:6-1966.

XVIII. Deformation stresses (in case of steel bridges only) Ref. Clause 219 of IRC:6-1966.

Live Load (Reference to the accountability of this force to bridge structures is Clause 207, 208 & 210 of IRC:6-1966)

Need for Standardising Live Loadings for Bridges

Highway bridges are required to be designed to support all vehicular loads safely that might move over it in its lifetime. It is not possible for the designer to know what vehicles will use the structure or what the required life of the bridge will be. This is to counter this problem and also ensure the safety of the structure, some form of control is required to be maintained and provisions formulated for sufficient strength to carry present and predicted future loads. The regulation of vehicles must be such that those of excessive weight are prohibited from crossing. Design control as concerned to bridges in India is provided by the Indian Roads Congress, which specifies the acceptable live loads. The traffic regulations covered by the State law specify the legal weight of motor vehicles. Until the nineteenth century, the live loads for the design of bridges were not a problem. The crossing over traffic was generally very light and not of any real significance. But with passing of time, the live loads became heavier, which called for the need to construct bridges capable of sustaining such loads. This resulted in the practice of specifying live loads for the design of highway bridges. The early loading standards were generally specified by local authorities based on the traffic which was likely to use the bridge facility. These loading were generally in the form of about 12 tonnes steam rollers.

The specification of a standard loading for bridges to cater the need of military transport and its heavy equipment was realised during the first world war. In 1972, UK introduced for the first time a standard loading train. The technological advancements and industrial progress led IRC to the development of some sort of standard loading for the design of highways bridges in India. IRC has evolved standards for bridge commensurate with traffic needs to Indian Highway system. These loading where introduced first time in India in 1939 and at that time were of two types - one known as IRC standard loading and other IRC heavy loading both consisting of distributed load and knife edge load.

The IRC standard loading consisted of a uniformly distributed load of 1.13 tonnes per linear metre of each traffic lane plus a knife edge load of 6 tons for computing bending moment and 9 tons for computing shear force. The IRC heavy loading was similar to the first, the only difference being the uniformly distributed load was increased by 0.8 tonnes/m and the knife edge loads were increased by 1.0 tons each. In 1958 these loading were replaced by a set of wheel load trains known as the IRC Class AA, Class A and Class B loading. In 1966, an additional loading known as class 70 R was introduced by bringing certain modifications in the already existing Class AA, hence the ' R ' added, meaning revised. Depending upon the type of bridge, the class of loading to be applied during analysis and design is chosen. It has been found that there is little relation between the class AA, Class A and B loading to the vehicles currently in use in the country. Class 70 R and Class AA loading specify a 70 tonnes tracked vehicle with only slight differences in the length of the loaded area. Thus although the vehicles are practically identical with the same total load, the minimum spacing between vehicles specified for the two load classes are very different - for class 70 R it is 30 m and for class AA it is 90 m.

It has been established that these classes of the IRC loading are critical for both bending and shear in different ranges of spans, Hence all these are usually tried in any bridge design.

IRC Class AA Loading

The IRC Class AA loading corresponds to the Class 70 loading and is based on the original classification methods of the Defence Authorities. This loading is to be adopted for design of bridges within certain municipal limits, in certain existing or contemplated industrial area, in other specified areas and along National Highway and State Highways. Two types of vehicles are specified :

i. Tracked Vehicles

ii. Wheeled Vehicles.

This loading consists of tracked vehicle, the loading of which is 70 tonnes and a wheeled vehicle of 100 tonnes loading. Both the vehicle types have specified dimensions which are to be observed during the live load analysis in bridge design. The choice from the two types is made depending upon the anticipated types of vehicles to travel on the bridge. Bridges designed for Class AA loading should be checked for class A loading also. As under certain conditions heavier stress may be obtained under class A loading.

IRC Class A Loading

The Class A loading was proposed by IRC with the objective of covering the worst combination of axle loads and axle spacing likely to arise from the various types of vehicles that are normally expected to use the road. This load train is reported to have been arrived at after an exhaustive analysis of all lorries made in all the countries of the world. The loading consists of a train of wheel loads that is composed of a driving vehicle and two trailers of specified axle spacings and loads. This loading in bridge designing is generally adopted on all roads on which permanent bridges and culverts are constructed.

IRC Class B Loading

The IRC Class B loading is similar to class A train of vehicles with reduced axle loads and tyre contact dimensions. This loading is to be normally adopted for temporary structure and for bridges in specified areas. Structures with timber spans are regarded as temporary structures. Class B loading is 60% of class loading. The positions of wheels and axle are same for both Class A and Class B loading.

IRC Class 70R Loading

Apart from these, the IRC has also given a load classification chart in Appendix I of IRC Standard Specifications and Code of Practice for Road Bridge. Section II (IRC 6) gives loading varying from class 3 to 70 R. The new class 70 R loading given in this chart is nothing but a revision of the Class 70 loading of the original classification i.e. the Class AA loading, with incorporation of certain changes mainly in case of wheeled vehicles which in the latest loading consists of a 100 tonnes trailer combination. With the introduction of this revised load classification, the road authorities in the country have prescribed this new class 70 R loading also for road bridges on all important routes such as National Highways. For multilane bridges and culverts, one train of Class 70R tracked or wheeled vehicles whichever creates severer conditions should be considered for every two traffic lane widths. No other live load should be considered on any part of the said 2 lane width carriage way of the bridge when the above mentioned train vehicle is crossing the bridge.

General Points to be Taken into Account during live Load Analysis for Bridge Design

During carrying out live load analysis for bridge design, the above standard loadings are to be placed and aligned along the length of the bridge between the kerb to kerb width of the carriageway, in such a manner so as to produce maximum value of bending moment or shear force at a particular section of a bridge member. This is acceptable, provided that the maximum clearances between a vehicle and the roadway face of a kerb and between two passing or crossing vehicles are not encroached upon. For each standard vehicle or train all the axles of a unit of vehicles should be considered as acting simultaneously in a position causing maximum stresses. Vehicles in adjacent lanes should be taken as headed in the direction producing maximum stresses. The spaces on the carriageway left uncovered by the standard train or vehicle should not be assumed as subject to any additional live load.

Review of the above Standard Loadings

Unlike India, most of the countries, particularly those leading in highway transportation, such as USA, Japan, UK and Germany do not specify hypothetical vehicles for their bridge designs but adopt more realistic heavy truck loads, in correspondence to those actually using the roads, as the loading train for the design of bridges. In some cases they also check important routes like the Inter-State Highways for the military loading which also is not very heavy.

Studies have shown that the IRC class A, class AA and class 70 R loadings, now specified for bridges on all important highways in India, are significantly heavier than those specified for similar bridges in many other countries. If economy should be aimed at in bridge projects in India, an urgent need seems to have arisen to make an immediate review of the present loadings with the object of making a suitable revision of these standards to make them more realistic to the traffic needs. The IRC bridge loadings are so complicated that the design of every bridge structure becomes a new problem involving labourious and tedious computations. Apart from these loading difficulties encountered in the effort to simulate these loading conditions both in the laboratories for experimental research work and in the field while carrying out full scale field tests, also perhaps call for a simplification of these loadings.

Reduction in Live Load Intensity

Where maximum stresses are produced in any member by loading more than one traffic lane simultaneously (as is generally the case), the following percentages of the resultant live load effects should be used in view of the improbably coincident number of lanes loaded :

One or two lanes	100%
Three Lanes	90%
Four lanes or more	75%

The reduction in intensity of floor beam loads is to be determined as in the case of main trusses or girders, using the width of roadway which must be loaded to produce maximum stresses in the floor beam.

Impact Effect of Live Load

Importance of Impact effect Due to Live Load on Bridge Design

A vehicle moving across a bridge at a normal rate of speed produces greater stresses than a vehicle that remains in a static position on the structure. This increment in stress can be called as the stress due to dynamic action of the moving load, which in bridge designing terminology is referred to as impact. Thus it can be said that impact factor is equivalent static effect of loading.

The total dynamic effect is felt not only when the wheel of a vehicles strike deck imperfections but also when the live load is applied to the structure for a very short time. As per the theory of dynamics a load instantly applied to a beam a causes stresses of twice the magnitude obtained when the same load is static on the beam. In actual load applications to highway bridges the total live loading is never instantaneous but is applied to the structure in a finite period. The dynamic effect due to sudden loading is variable for all structural elements of a bridge. In addition to the true impact and sudden loading loading effects, a third effect is caused by the vehicle vibrating on its spring. Uneven roadway surfaces contribute to this effect. The vibration of the vehicles induces vibrations in the structure.

The magnitude of stresses is dependent on the relative masses of the vehicle and the bridge, the natural frequency of the structure, and the damping characteristics of the bridge. It is observed that the impact factors for concrete bridges are less than those of steel bridges. This is so because the damping properties of concrete are better as compared to that of steel.

From the above discussion it can be conclusively said that impact factor is dependant on the following parametres as follows:

a. Span length of the bridge : With increase in span length the value of impact decreases.

b. Speed of vehicles : An increase in the speed of the live load results in an increase in the value of the impact factor.

c. Bridge material used : Impact factor decreases with an increase in damping. Due to this reason impact factors for concrete bridges are less than those of steel bridges.

d. Condition of expansion joint at the beginning of the bridge : Impact factor is less in case the expansion joint at the beginning of the bridge is smooth and vice - versa.

e. Condition of road surface : Smoother roads lead to lower values of impact factor and vice - versa.

f. Natural frequency of the bridge structure & the vehicle moving on it: This is so because in case resonance occurs, then this will result in an increase in the value of impact factor. It has been observed that percentage wise increase in impact factor due to the additive effect of moving vehicles on vibration of the bridge structure will be more for small span bridges rather than for longer spans ones. Also the chances of the coinciding of the damping at the centre and the moving vehicle being also at that point is more for small bridges. These chances are enhanced at higher speeds of the vehicles.

The provision for impact effect in bridge designing is made by an increment of live load by an impact allowance expressed as a fraction or a percentage of the applied live load. This allowance varies for different loadings eg. : In case of Class A & Class B loading the value of impact factor is :

i. For reinforced concrete bridges, impact factor = 4.5/(6+L)

ii. For steel bridges, impact factor = 9/(13.5+L)

It is to be noted that the effect of impact is considered only while designing the superstructure, bearings, abutment or pier caps. For designing the substructure this effect is not considered.

Longitudinal Forces

Effects of Horizontal Forces & Externally Applied and Self Induced

The bridge is to be designed to account for the effects of the extremely applied horizontal forces like braking which is generated by the application of brakes on the live load and self induced horizontal forces usually caused by movements/rotation of the deck due to temperature variation, shrinkage and creep of deck concrete and due to elastic shortening of the deck due to prestress. The method for evaluating the horizontal forces caused at the various bridge supports as a result of the distribution of these forces as effected by the type of bearings, shear ratings of the supports, and location of the zero movement point in the deck is by provision for longitudinal forces arising from any one or more of the following cases in all road bridges:

a. Tractive efforts caused through acceleration of the driving wheels.

b. Braking effect resulting from the application of the brakes to braked wheels. Braking effect is invariably greater than tractive efforts.

c. Frictional resistances offered to the movement of free bearings due to change in temperature or any other cause.

The braking effect on a simply supported span or a continuous unit of spans or of any other type of bridge unit should be assumed to have the following value :

- In the case of a single lane or a two lane bridge : Twenty percent of the first train load plus ten percent of the load of the succeeding trains or part thereof, the train loads in one lane only being considered for the purposes of the sub clause. Where the entire first train is not on the full span, the braking force should be taken as equal to twenty percent of the loads actually on the span.
- In the case of Bridges, having more than two lanes : As in (1) above, the first two lanes plus five percent of the loads on the lanes in excess of two.

The above loads are not be increased on account of impact while designing. The force due to braking effects is taken to be acting along a line parallel to the carriageway and 1.20 metre above it. Due to this reason, while the force is being transferred to the bearing for design analysis, the change in the vertical reaction at the bearings is to be taken into account.

The longitudinal force at any free bearing should be limited to the sum of dead and live load reactions at the bearing multiplied by the appropriate coefficient of friction. The coefficient of friction at the bearing should be assumed to have the following values e.g. :

- For roller bearings 0.03
- For sliding bearings of hard copper alloy 0.15
- For sliding bearing of steel on cast iron on steel 0.25

The longitudinal force at the fixed bearing should be taken as the algebraic sum of the longitudinal forces at the free bearings in the bridge unit under consideration and the force due to braking effect on the wheels as mentioned above.

The effect of braking force on bridge structures with bearings such as arches, rigid frames, etc. should be calculated in accordance with approved methods of analysis of indeterminate structure.

The effect of the longitudinal forces and all other horizontal forces should be calculated upto a level where the resultant passive earth resistance of the soil below the deepest scour level (floor level in case of bridge having pucca floor) balance these forces.

Self induced horizontal force is caused at bearings by movement/ rotation of deck due to temperature variation, creep and shrinkage of deck, concrete elastic-shortening of deck due to prestress etc.) Ref. Clause 218 of IRC:6-1966.

Wind Load

Wind effects the bridge structures directly or indirectly. By its direct action on vehicular loads, pedestrians and the structure itself, wind poses serious problems due to resultant vibrations specially in case of cantilever type bridges, suspension bridges which are exposed. The indirect effects of the wind force arises from the feeling of instability it arouses within the bridge user due to excitation of the structure caused by it. These forces are considered to be acting horizontally and in such a direction that the values of resultant stresses are maximum for the bridge member under consideration.

The wind force on a bridge structure is treated as a horizontal force of specified intensity and is taken to be acting on an area calculated in the following manner :

- In case of bridge superstructure : For this the area of the superstructure as seen in elevation along with the floor system and railings is considered. In case of railings and parapet wall having perforations, the resultant area is obtained by deducting area covered by perforations from the above calculated area.
- In case of through or half through type superstructure : For this the area of the elevation of the windward truss as specified in (a) above along with half the area of elevation above the deck level of all other trusses or girders is considered.

The force due to wind acting laterally on any exposed moving live load is taken to be acting at a height of 1.5 m above the formation level of the bridge. This value is taken to be 300 kg/m in case of highway bridges with ordinary moving live loads and 450 kg/m for bridges with tramway loads. While calculating the total force coming on a bridge due to wind, the clear distance between successive trains of live loads is also considered. When the wind velocity at deck level is above 130 km/ hour the bridge is considered to be carrying no live load.

On through or half through truss latticed or other similar spans the total assumed wind force is taken to be less than 450 kg /m in the plane of the loaded chord and 225 kg /m in the plane of the unloaded chord, and not less than 450 kg /m on deck spans.

In case a wind pressure of 240 kg / m2 of the unloaded structure results in greater stresses than those produced by the combined wind forces as stated above, then this should be considered in the design.

Usually from design point of view the wind forces discussed above are treated as static, a treatment that is usually satisfactory when the structure is rigid. But for flexible structures this procedure is not sufficient. Suspended structures involve a more complex procedure. A steady state wind may set up an effect known as Vortex Shedding when the structure has a long period of natural frequency and a minimum of natural damping. The phenomenon of vortex shedding involves production of vortices as the wind passes around the structures, as a consequence of which pressures and sanctions alternate on each side of the obstruction to the wind. This alternating direction of forces at a right angle to the wind causes the structure to vibrate in a direction perpendicular to natural period of the structure with amplitudes and vibrations large enough to create damage to the bridge. For development of resonance in case of stiff structures the wind velocity will have to be greater than normal but for flexible structures, however, resonance can occur at even low values of wind velocities.

Seismic Force

Earlier, while designing bridges the effects of earthquakes on highway bridges were generally ignored or given consideration only with regard to the design of substructures. Superstructures usually have adequate strength to resist any inertia force in the vertical or horizontal direction. However, seismic failures of several bridges in recent years have shown that the failure was not caused by the collapse of any element of the superstructure but rather as a consequence of :

(1) the superstructure shaking off the bearings and falling to the ground

(2) the structural failure due to the loss of strength of the soil under the substructure as a result of the vibrations induced in the ground. During the event of an earthquakes it is important that bridges remain in place to allow mobility of relief supplies. This becomes evident from examples of the catastrophe created due to severe earthquakes in San Francisco, Mexico and Japan. The common factor which resulted in such disasters in these areas was the poor soil quality on which structures were built.

The influence of seismic forces on a structure depends on the bridge's elastic characteristics and the distribution of weight. A rigorous analysis is complex and involves the application of structural dynamics. In addition, it is necessary to establish the ground motion expected under the sub structure. If a bridge is situated in a region subjected to earthquakes allowance should be made in the design for the seismic force by introducing earthquake resistant features in the bridge.

The seismic force for calculation purpose in the design is taken as a horizontal force equal to the appropriate fraction of the weight of the dead and live load acting above the section under consideration. Parts of the structures embedded in soil are not be considered to producing any horizontal forces. As per IS 1893-1970 seismic force to be used in the design of a structure is dependent on many variable factors and therefore, it is extremely difficult to determine its correct value. To give broad indications of reasonable values of seismic coefficient for different regions of India, India has been categorised into in to five seismic zones designated as zones I to V. The degree of severity of seismic forces increases from zone I to zone V.

The seismic coefficients for horizontal forces are given in the code as per the zone and also depending upon the foundations soil which has been assumed to be hard, medium and soft type. The vertical seismic coefficient where applicable may be taken as half of the horizontal seismic coefficient for design purposes.

Estimation of Seismic Forces for Live Load

The seismic force due to live load can be ignored when acting in the direction of the traffic i.e, in the longitudinal direction of the bridge, but should be taken into account when acting in the direction perpendicular to traffic i.e. in the transverse direction of the bridge. In the latter case, the seismic force is to be computed for 50% of the design live load for railway bridges and 25% of the design live load for road bridges as specified in the relevant standards. The superstructure should be designed to resist horizontal and vertical seismic force.

In case of stability analysis for foundations against overturning in the transverse direction, a factor of safety of 1.5 is assumed in the design. The bridges built in zones 4th and 5th, should have a special provision to secure the superstructure to the piers to prevent it from being dislodged off its bearing during an earthquake. Masonry arch bridges of spans more than 10 m should be avoided third to fifth zones.

For the design of substructure the following seismic forces are considered above the scour depth :

- Horizontal and vertical forces due to dead, live and seismic loads transferred from superstructure to the substructure through the bearings.

- Horizontal and vertical seismic forces due to self weight applied at the centre of mass ignoring reduction due to buoyancy or uplift.
- Hydrodynamic force acting on piers and modified earth pressure on abutments due to earthquake.

The above seismic forces are assumed to act parallel to the traffic directions whichever produces the severe effect in the design of piers. In case of portions of the piers which are submerged, the hydrodynamic force F assumed to act in a horizontal direction corresponding to that of earthquake motions is calculated as by using the following equation :

$$F = A*a*W,$$

here A = a coefficient which depends upon the ratio of height of submerged portion of pier to its radius.

a = horizontal seismic coefficient.

W = weight of water surrounding the pier.

In the design of substructure the reduction in weight due to buoyancy is not taken into account while calculating the seismic force. Also the bridge structure is with the assumption that the seismic and wind forces are not acting simultaneously and similarly the maximum seismic force and highest flood case are considered to be occurring simultaneously.

Flood Water Current Forces

Flood Water force on Substructure

While designing bridges if any portion of the structure is submerged in flowing water then it is to be also designed to safely resist the horizontal force due to the water current. Importantly bridge substructures constructed in a region of flowing water should be designed to withstand water pressure which could cause the pier to slide or overturn. Scour around the bottom of the piers is of great concern in the bridge design. The design for every bridge over a stream should involve a careful study of possible stream velocities. Past flow records are helpful to determine velocities.

If this information is lacking, the design engineer should make the best estimate possible with the available data. Any part of a bridge which may be submerged in running water should be designed to sustain safely the horizontal pressure due to the force of the current. In case of piers parallel to the direction of water current, the intensity of pressure should be calculated from the following formula :

$P = 52\ KV^2$

Here, P = Intensity of pressure in kg/m^2 due to the water current.

K = a coefficient whose value depends upon the different shapes of the piers eg.

Sr.No.	*Shape of Piers, K value*
1	Square ended piers, 1.50
2	Circular piers or piers with semi circular ends, 0.66
3	Piers with triangular cut and case waters, the angle included between the faces being 30 degrees or less, 0.50

V = the velocity of the current in metres/sec. at the point where the pressure intensity is to be computed The value of V^2 in the formula given above should be assumed to vary linearly from zero at the point of maximum scour to the square of the maximum velocity at the free surface of water. The maximum velocity for this purpose is considered as 2 times the maximum mean velocity of the current.

When the current strikes the piers at an angle, the velocity of the current should be resolved into two components one parallel and the other normal to the pier.

Thus in order to account for any possible variations in the direction of the current from the direction assumed in the design, allowance should be made in the design of piers for an extra variation in the current direction of 20 degrees. The flood current, flowing at a maximum velocity V m/s or a maximum mean velocity v m/s may be assumed to strike the substructure at angle, q = 20 to the normal flow direction, so that flood strikes the width of pier @ V cos. q m/s and the length of pier @ V sin q m/s.

The resulting dynamic stream pressures on the substructure in the two orthogonal directions respectively, may be assumed to vary linearly from zero at maximum scour level to the maximum of 52 K $(V \cos q)^2$ and 52 K $(V \sin q)^2$ kg/m^2 at the high flood level. V is taken equal to 2 v. V and v in m/s. This dynamic water pressure triangle is unlike the static pressure triangle which has a variation of being zero at the surface to maximum at the bottom of water depth.

Apart from the above calculations, depending upon the design requirements the following may be considered also :

I. Effect of afflux head which is created locally on a pier as a result of the flood flow being obstructed by the pier.

II. Effect of crosscurrent force in longitudinal direction of the bridge is considered when the river bed is unerodible (rock) at scour level. The effect of about 25 cm static head of water across the pier faces looking at the bridge length is considered for design purposes. This force is usually taken as an alternative to the flood water force in bridge longitudinal direction.

In snow bound areas floating ice generally causes high forces to work against piers. In some cases bridges have been completely demolished by the pressure of floating ice. According to the AASHTO specifications a pressure of 30 kg/m2 of contact area of ice and substructure is taken in the design if required. The thickness of the ice and its point of application on the piers is to be determined by the engineer through proper investigations.

Buoyancy Effect (Floatation)

In case of high level bridges to allow for full buoyancy a reduction should be made in the gross weight of the member affected in the following manner for the purpose of design :

a. When the member under consideration displaces water only e.g. a shallow pier or abutment founded at or near the bed level, the reduction in weight should be equal to that of the volume of the displaced water.

b. When the member under consideration displaces water and also silt or sand e.g. a deep pier or abutment passing through strata of sand and silt and founded on similar material, the upward pressure causing the reduction in weight should be considered as made up of the following two factors :

(1) Full hydrostatic pressure due to a depth of water equal to difference in levels between the free surface of water and the foundation of member under consideration, the free surface being taken for the worst condition.

(2) Upward pressure due to the submerged height weight of the silt or sand calculated in accordance with Rankine's theory for the appropriate angle of internal friction.

In the design of submerged masonry or concrete structures, the buoyancy effect through pore pressure may be limited to 15 per cent of full buoyancy.

In case of submersible bridges, the full buoyancy effect of the superstructure, piers and abutments should be taken into consideration.

Erection Stresses

Construction, Handling and Erection Loads

Erection Stresses : Erection stresses are the stresses set up in any member during erection, temporary erection conditions and partial span-dislodgment conditions. These stresses are different from those occurring when the structure is in position and in actual service. They should be checked, if unsafe, provision should be made for them. While designing consideration should be given to the effect of temporary loads imposed by sequence of construction stages, forming falsework and construction equipment and the stresses created by lifting of placing precast members. Generally before execution, a construction scheme is considered for the feasibility of the project and the contractor is usually left to his own ingenuity for developing the construction procedure. The stability of precast members during and after construction should be investigated and provisions made for stability as needed. Effects of member-shortening and redistribution of loads during prestressing should be considered. Owing to long term creep effect the distribution of moments, shears, etc. in a continuous structure initially built in parts with different span arrangement for each casting becomes nearly same as if the whole structure was cast monolithically in one go. This requires for a workmanlike understanding of true stress build up rather than only a mere numerical superimposition of stresses.

Thus from above it can be seen that whole or parts of a bridge may be subjected to several loads simultaneously. The engineer has to decide what combinations of loads and magnitudes will most likely be applied at one time. For some combinations the normal allowable stress can be increased, which is an indirect way of saying that only a percentage of the sum of the maximum effects is considered because the possibility that the peak value of all these events occurring simultaneously is remote. As an example, when dead load, earth pressures, effect of buoyancy, stream flow, and wind loads are applied simultaneously the allowable stresses can taken to be 125% of the normal allowable stresses. This amounts to say that only 80% of the peak values of these loads would be considered as acting together.

Deformation Stresses

These stresses are considered only in the case of designing steel bridges. A deformation stress is actually a bending stress in any member of an open web girder or a truss, caused by the vertical deflection of the girder combined with the rigidity of the joints. All steel bridges are designed and erected in such a manner that these stresses are reduced to a minimum.

In the absence of any data or calculation, these stresses should be assumed to be not less than 16% of the dead load and live load stresses. In prestressed steel girder bridges, deformation stresses may be ignored.

Importance of Bearings

Bearings play an important role in the design of bridges. Bearings transmit the load coming from the superstructure to the substructure. These also allow for specific movement of the superstructures namely :

- Temperature variations related to longitudinal movement of the deck
- Rotation as a consequence of deck deflections
- Vertical movements as a result of sinking of any of the supports
- Movements due to creep, shrinkage or prestressing which cause elastic shortening
- Movements due to applied horizontal loads on the superstructure can be either transverse or longitudinal or can be coexisting, with respect to the bridge centre line.

Loads transverse to bridge centre line : Wind or seismic effects on both the bridge deck and on the traffic using the bridge, and also the centrifugal force from traffic using the bridge in case of the deck being curved.

Loads parallel (longitudinal) to the bridge centre line : In addition to the effects of wind or seismic forces, there is also the braking force which is usually taken as a percentage of the live load coming on the bridge.

Description of Bridge Deck Movements

The horizontal movements of a bridge superstructure are due to:

I. Temperature expansion and contraction

II. Shrinkage of concrete

III. Shortening of concrete due to creep effect

IV. Elastic shortening

V. Movements due to induced external loads (e.g. earthquake, wind, vehicular braking etc.)

Temperature

Temperature variations cause both expansion and shortening of the bridge deck and are usually computed as a plus and minus range about a mean structure temperature which occurs when the superstructure is placed on the bearings or as per the codal recommendations. Temperature differentials also occur in the deck from top to bottom and from one side to other side of the deck. Temperature differentials through the depth of the deck have little effect on the bearings and piers but those from one side of the deck to the other cause the deck to bend in plan which results in horizontal forces on the bearings and piers.

Shrinkage of Concrete

Shrinkage of concrete results in the shortening of the bridge superstructure. This effect depends upon factors such as quality of concrete used, size of the member, relative humidity and time after casting.

Shortening due to Creep Effect

Effect of creep of concrete under prestressing and other permanent loads results in the shortening of the superstructure. It is a time dependent effect.

Elastic Shortening

This phenomenon occurs in case of prestressed superstructure during prestressing. The amount of shortening depends on the stage at which the superstructure is placed on the bearings and also at what stage it is prestressed. Some times, partial or complete prestressing may be carried out before the superstructure is placed on the bearings thus eliminating this shortening from affecting the bearings at least partly.

Bearings for Bridges

Types of Bearings Recommended for Various Spans - Lengths and Support - Flexibility Conditions

Bridge bearings must be designed to transmit all vertical loads and appropriate horizontal forces (depending on the functionality of the individual bearings).

i. Generally where a simple span deck is supported over rigid supports like in case of solid slab deck (unlike the adjoining cantilever tips in segmental free-cantilever construction, which are flexible) and the span is less than about 7.5 m, no special bearings are necessary - only tar paper or a felt layer is adequate. Here the concrete surfaces which are going to be in contact are smoothened by carborundum stone).

ii. For spans between about 7.5 m to about 15 m, mild steel plate bearings, sliding type over free supports and rocking type (with a rocking pin) over the fixed support, may be used. Alternatively, neoprene or elastomeric bearings (pads or strips or laminated type) may also be used.

iii. For spans exceeding 15 m, metallic rocker type bearing is provided over the fixed support and roller cum rocker type bearing is provided over the free support. Alternatively, suitably designed laminated neoprene or elastomeric bearings may be provided. Usually, neoprene or elastomeric bearings are provided upto span length of 30 m.

Note : Even for short simple span decks, placed across flexible cantilever tips mentioned in (i) above full fledged rocker and roller rocker metallic bearings (described in (iii) above) are advisable because of considerable deck movement and deflection and (consequent rotation) at the tips of the long cantilevers. Alternative types of bearings may be used, but these normally would not include the laminated neoprene bearings if the vertical reactions low and deformations high.

Structural Aspects to Various Types of Bearings

Elastomer : A compound containing natural or chloroprene rubber with properties similar to those of rubber.

Roller bearing: A bearing consisting essentially of one or more steel rollers between parallel upper and lower steel plates.

Rocker bearing: A bearing consisting essentially of a curved surface in contact with a flat or curved surface and constrained to prevent relative horizontal movement. The curved surfaces may be cylindrical or spherical. Rocker bearings permit rotation by rolling of one part on another.

Knuckle bearing: A bearing consisting essentially of two or more members with mating curved surfaced. The curved surfaces may be cylindrical or spherical. Knuckle bearings permit rotation by sliding of one part on another.

Leaf bearing: A bearing consisting essentially of a pin passing through a number of interleaved plates fixed alternatively to the upper and lower outer bearing plates.

Sliding bearing: A bearing consisting essentially of two surfaced sliding one on the other.

Elastomeric bearing: A bearing comprising of a block of elastomer that may be reinforced internally with steel plates (steel laminated) - which make it a laminated or restrained elastomeric bearing.

Plain and bearing: An unreinforced elastomeric bearing

Strip bearing: A plain pad bearing for which the length is at least ten times the width.

Pot bearing: A bearing consisting essentially of a metal piston supported by a disc of unreinforced elastomer that is confined within a metal cylinder or pot.

Expansion Joints for Bridge Superstructures

Movements at Joints as Compared to those Taken up by the Bearings

As discussed above, the movements to be catered for at bearings locations and expansion joints arise from the following causes :

i. Thermal expansion and contraction of the superstructure and in certain cases even of substructure
ii. Shrinkage of concrete
iii. Creep in the concrete
iv. Elastic shortening under prestress
v. Displacement of the structure under load or any other action.

Because expansion joints are installed at a late stage in construction, therefore, some of these movements will already at least partly have been taken place. Hence less total movement has to be catered for at expansion joints than in bearings at the same location.

Elastic shortening under pressers is normally assessed on the mean stress included in the deck section by the prestress and dead loads. In most bridge, displacements of the structure under load produce very minor movements.

Also regarding the movement at the expansion bearing, for various expanding lengths in case of a cast in situ prestressed concrete deck, it has been observed that the movement at the bearing is more than at the expansion joint for the same deck since the latter is installed much after the deck is cast on top of the bearing and therefore, suffers movements only due to temperature and parts of concrete shrinkage and creep.

Design Philosophy in Case of Expansion Joints

The basic philosophy usually adopted is that minimum number of joints be used consistent with the need for articulation of the structure. The approach to eliminate as may expansion joints as possible during the designing stage is taken because of the generally poor performance of deck joints. There are advantages to be derived from eliminating joints entirely, where possible, as in the case for curved structures or those which have flexibility in the support system, such as with the slender or hinged columns. The structure acts like a bow with the superstructure fixed at both abutments. Movement at intermediate support is accommodated by sliding bearings.

Where deck joints are required the following two types should be generally excluded because of unsatisfactory service records; any joint which has been found prone to failure within a few years of installation and joints placed directly in concrete without provision for armour at its edges with the roadway surface as a consequence of which they fail because of spelling of the concrete. This type of failure was common when it was the normal practice to install compression seals in direct contact with a concrete deck slab.

Deck joints divide into two categories, one of which are those which are designed to be sealed (waterproof) and the other which are of the open type (designed to permit the fare flow of water through the joint). All joints must be designed so that none of the components fail at serviceability limit stages and the sealed joints hold remain watertight to prevent water leakages. Since deck joints do not directly affect the safety of a structure, some damage can be tolerated from conditions exceeding serviceability limit states because it is uneconomical to design joints for ultimate limit states. Damage which is tolerable is that damage which can be repaired while the deck remains in service, even though one or more lanes may have to be closed to traffic to permit the repairs to be carried out.

Sources Leading to Movements and Loads at Expansion Joints

Movement in a structure result from either loading conditions or due to change in volume of the components of the structure. The magnitude of the movement while designing is to be calculated for each structure since it depends upon the nature of the structure and its response to static and dynamic forces. The most common factor resulting in loads on and movements in deck joints are direct impact of wheels, temperature and those caused by soil conditions which result in the movement of supports. Conversely some joints particularly compression seals or those which accommodate movement by the shear deformation of an elastomer can impose some forces on the structure which may have to be included in the design of the structure. This characteristic of compression seal and certain other joints may prevent replacement of these joints for those which do not exert forces on the structure - unless the forces are not significant.

Following are the factors affecting the magnitude of movements and loads that have to be considered in the design as given below :

a. Properties of the materials used for the bridge structure including constants such as coefficient of thermal expansion, modulus of elasticity, Poisson's ratio, creep and shrinkage strains
b. Effective temperature variations in the structure
c. Sizes and properties of the structural members
d. Method and sequence of construction. This becomes critical when construction and loadings are done in stages
e. Tilt, settlement and movement of supports
f. Construction tolerances
g. Static and dynamic response of the bridge structure
h. Interaction of the forces to which the structure may be subjected including dead and live loading; wind, seismic effects and earth pressures
i. Structural restraints, if any.

Joints which are considered 'fixed' may be fixed against translation, but not rotation. Flexural deflection of a member results in rotation at the joint and failure to accommodate this rotation has been observed to be a common cause of water leakage and deterioration at fixed joints.

The plan geometry of a bridge superstructure is an important factor in determining the movements which take place at different points in the deck structure. Bridges which are straight and narrow in plan normally experience movements along their centreline in response to changes in temperature. Bridges which are curved in plan or with skew joints have two components of movement at deck joints, a longitudinal component along the centreline of the superstructure and a transverse component perpendicular to the longitudinal centreline. As a consequence of the transverse component serious problems with some types of expansion devices are caused, for example finger joints, sometimes also called tooth or comb joints. Binding up of the joint can result in large forces on the joints, bearings and abutments with corresponding reactions at intermediate piers.

Response of Curved Structures

To determine the actual movement of a bridge superstructure at all the points of attachment of the superstructure the points of the zero movement point have to be established. The position of this point depends upon the geometry of the superstructure the location of the piers and bearings and their relative shear stiffness. For straight structures the zero movement joint is generally located on the longitudinal centreline of symmetry. In curved structure the zero movement point will generally not be located on the longitudinal centreline and may even fall outside the plan of the superstructure. Movements are always towards and away from the zero movement point. The calculations of the location of the zero movement point can be quite involved for complex structure.

Types of Expansion Joints

Open type Joints

An open type of joint can provide a good enough, effective and an economic solution under certain conditions. Satisfactory performance depends upon the provision of an effective deck drainage system, control of flow of water, of surface drainage through the joint and disposal of the runoff from the site. The success of open joints in the past has been limited by the damage to the portion of the structure immediately beneath the joint from debris and run off. Wet dirt often accumulates on the flanges of steel beams or around bearings, and leads to corrosion. Chlorides salts penetrate concrete surface, thereby causing the spelling of concrete and the corrosion of embedded reinforcement. The amount of deterioration can be kept to a minimum by removing the accumulated debris by routine washing, followed by spot repairing in steel structures.

Sealed Type Joints

Sealed type of joints are required to seal the surface of the deck, including kerbs, sidewalks, medians and even the barriers or parapet walls, against the seepage of moisture and other contaminants through the joint. While designing care should be taken so as to make sure that the geometry of the joint be such as to prevent the accumulation of water and roadway debris in the joint. Metal to metal sliding surfaces have not performed satisfactorily, largely because of their effects of vibrations the accumulation of debris and the fact that they cannot be sealed effectively. The presence of debris and in the winter months, ice has been found to restrict the movement capabilities of joint, displacing thin membrane seals. Where the movement of the joint is accommodated through the change in geometry of a thin membrane strip, these should not be in direct contact with vehicles in order to prevent their failure from wear and puncture by the trapped incompressible. Some of the preparatory joints, claimed to be 'self cleansing' have been found to be no less vulnerable to damage by accumulated debris than other types of joints. Deck joints having elastomeric components in direct contact with the deck surface, do not provide adequate support for the adjacent edges of the deck wearing surface. Nosings of epoxy mortal or similar material are sometimes used between the edges of the deck surface and the joint to try and prevent spelling of the concrete edges of the deck.

The chief factors in determining whether to use an open or a sealed point are the cost and the location of the site. The total costs of sealed joints may include the cost of providing for additional forces on the structure, and an increasing in maintenance costs and frequency of inspection. In general open joints are well suited to secondary highways where little sand and salt is supplied during winter maintenance operations. Sealed joints tend to be more suited to urban areas where the costs of providing for the surface run off can be high.

Elastomeric (Neoprene) Compressed Seals

There are two basic types of extruded elastomeric seals. One is the open cell compression seal dependent upon its ability to maintain pressure on the joint side walls with varying degrees of stress. The other type consists of strip seals and box configuration seals using locking lugs or ears placed in compression within the cavity of a retainer. The former seals may be designated compression seals and the latter addressed as either strip seals or box seals for whatever case applies as will be seen in the figures ahead. These types of seals are necessarily dependent upon the material they are made of. This material is that makes performed elastomeric seals superior to all other types of seals and sealants. Generically, the basic material is polychloroprene, otherwise known as Neoprene. Polychloroprne is used because of its agreeable physical properties. Since the first uses, of these seals were oriented to the transportation field, they had to resist deterioration from exposure to whether sunlight oils, chemicals, heat, abrasion, and impact, while still remaining elastic during their expected life cycle.

Compression Seals

Compression seals as the name suggests are seals that compressed when inserted into the joint opening and remain in the state of compression during all movement phases of the joint. They are generally designated by size according to their width. A movement rating is usually established for each width seal, along with other pertinent data by the manufacturer. These critical dimensions and data are available from the various manufacturers. The seal to be used in any particular case must have a nominal width which is greater than the largest expected gap opening. Therefore, the working dimensions of compression seals, vary generally from 85% of nominal width, when the joint is fully open to approximately 40 - 50 % of nominal width, when the joint gap is at its minimum dimension. Compression seals are always in stressed state. They maintain water tightness and keep out in compressible by maintaining side wall pressure on the joint interfaces.

Design Methodology for Compression Seal Joints

In order to design an expansion joint, the design engineer should first determine the joint movements that will take place due to temperature, creep, shrinkage, elastic shortening, etc. After this anticipated movement is ascertained, the joint geometry can easily be determined. The joint geometry or configuration for the entire movement cycle should accommodate the physical characteristics of the compression seal selected which implies that the physical properties of a seal exhibiting a movement rating equal to or grater than the expected joint movement.

After the obtaining the joint movements, these values in reference to the manufacturer charts are used for getting the maximum and minimum gap openings as well as other data for the seal selected. After getting the ambient temperature range, design can be proceeded with calculations for the joint geometry. Satisfactory performance of compression seals is judged upon by their ability to maintain side wall contact with the joint gap interfaces at all times, thereby creating a watertight seal and preventing in compressible from accumulating in the joint or penetrating downward along the side walls.

Therefore, several physical properties of a compression seal have been identified which serve to provide data to ascertain the expected performance capability of the elastomer and, hence, of the seal itself. These various parametres have been incorporated into specifications and certain test procedures have been developed for quality assurance.

Chapter 6

Construction Engineering Basics

Construction engineering is a professional discipline that deals with the designing, planning, construction, and management of infrastructures such as highways, bridges, airports, railroads, buildings, dams, and utilities. Construction Engineers are unique such that they are a cross between civil engineers and construction managers. Construction engineers learn the designing aspect much like civil engineers and construction site management functions much like construction managers.

***Figure** : The Berlin Brandenburg Airport under construction. The facility starts full operation in 2012.*

The primary difference between a construction engineer and a construction manager is that the construction engineer has the ability to sit for the Professional Engineer license (PE) whereas a construction manager cannot. At the educational level, construction managers are not as focused on design work as they are on construction procedures, methods, and people management. Their primary concern is to deliver a project on time, with in budget, and of the desired quality.

The difference between a construction engineer and civil engineer is only at the educational level as both disciplines are able to sit for the PE exam giving them the same title of engineer. Civil engineering students concentrate more on the design work, gearing them toward a career as a design professional. This essentially requires them to take a multitude of design courses. Construction engineering students take design courses as well as construction management courses. This allows them to understand both the design functions as well as the building requirements needed to design and build today's infrastructures.

Work Activities

Depending on which career path the construction engineer has chosen to follow, an entry-level design engineer normally provide support to project managers and assist with creating conceptual designs, scopes, and cost estimates for the planning and construction of approved projects. It should be noted that a career in design work does require a professional engineer license (PE). Individuals who pursue this career path are strongly advised to sit for the Engineer In Training exam (EIT) while in college as it takes five years post graduate to obtain the PE license. Entry-level construction manager positions are typical called project engineers or assistant project engineers. They are responsible for preparing purchasing requisitions, processing change orders, preparing monthly budgeting reports, and handling meeting minutes. The construction management position does not necessarily require a PE license; however possessing one does make the individual more marketable, as the PE license allows the individual to sign off on temporary structure designs.

Abilities

Construction engineers are problem solvers, they help create infrastructure that best meets the unique demands of its environment. They must be able to understand infrastructure life cycles and have the perspective to solve technical challenges with clarity and imagination. Therefore individuals should have a strong understanding of math and science, but many other skills are required, including critical and analytical thinking, time management, people management and good communication skills.

Educational Requirements

Individuals looking to obtain a construction engineering degree must first ensure that the program is accredited by EAC or Technology Accreditation Commission (TAC) of the Accreditation Board for Engineering and Technology (ABET). ABET accreditation is assurance that a college or university program meets the quality standards established by the profession for which it prepares its students. In the US there are currently twenty-five programs that exist in the entire country so careful college consideration is advised.

A typical construction engineering curriculum is a mixture of engineering mechanics, engineering design, construction management and general science and mathematics. This usually leads to a Bachelor of Science degree. The B.S. degree along with some design or construction experience is sufficient for most entry level positions. Graduate schools may be an option for those who want to go further in depth of the construction and engineering subjects taught at the undergraduate level. In most cases construction engineering graduates look to either civil engineering, engineering management, or business administration as a possible graduate degree. For authority to approve any final designs of public projects (and most any project), a construction engineer must have a professional engineers (P.E.) license.

Job Prospects

Job prospects for construction engineers generally have a strong cyclical variation. For example, starting in 2008 - continuing until at least 2011 - job prospects have been poor due to the collapse of housing bubbles in many parts of the world. This sharply reduced demand for construction, forced construction professionals towards infrastructure construction and therefore increased the competition faced by established and new construction engineers. This increased competition, and a core reduction in quantity demand is in parallel with a possible shift in the demand for construction engineers due to the automation of many engineering tasks, overall resulting in reduced prospects for construction engineers. In early 2010 the United States construction industry had a 27% unemployment rate, this is nearly three times higher than the 9.7% national average unemployment rate. The construction unemployment rate (including tradesmen) is comparable to the United States 1933 unemployment rate - the lowest point of the Great Depression - of 25%.

Remuneration

The average salary for a civil engineer in the UK depends on the sector, and more specifically the level of experience of the individual.

A 2010 survey of the remuneration and benefits of those occupying jobs in construction and the built environment industry showed that the average salary of a civil engineer in the UK is £29,582. The average salary varies depending on experience, for example the average annual salary for a civil engineer with between 3 and 6 years experience is £23,813. For those with between 14 and 20 years experience the average is £38,214.

Highway Engineering

***Figure** : Brazil's Rodovia Anchieta.*

Highway engineering is an engineering discipline branching from civil engineering which involves the design, construction and maintenance of Highway Systems. Highway Engineering become prominent towards the latter half of the 20th Century after World War 2. Standards of highway engineering are continuously being improved. Highway Engineers must take into account future traffic flows, design of highway intersections/interchanges, geometric alignment and design, highway pavement materials and design, structural design of pavement thickness, and pavement maintenance.

History of Highway Engineering

The beginning of road construction could be dated back to when the Romans were around. With the advancement of technology from carriages pulled by two horses to vehicles with power equivalent to 100 horses, road development had to follow suit. The construction of modern highways did not begin until the late 19th to early 20th century. The first research dedicated to highway engineering was initiated in the United Kingdom with the introduction of the Transport Research Laboratory (TRL), in 1930. In the USA, highway engineering became an important discipline with the passing of the Federal Aid Highway Act of 1944, which aimed to connect 90% of cities with a population of 50,000 or more. With constant stress from vehicles which grew larger as time passed, improvements to pavement were needed. With technology out of date, in 1958 the construction of the first motorway in Great Britain (the Preston bypass) played a major role in the development of new pavement technology.

Design policies standards used in the United States are typically based on publications of the American Association of State Highway and Transportation Officials as well as research promulgated by the Transportation Research Board, the Institute of Transportation Engineers, the Federal Highway Administration, and the Department of Transportation.

Bridge Engineering

Bridge is a structure which provides a passage to people, vehicles, railways or pipelines to cross various obstacles to travel. Engineers build bridges over obstacles such as lakes, rivers, canyons, and dangerous roads and railway tracks. Without bridges, people would need boats to cross waterways and would have to travel around canyons and ravines. Bridges range in length from a few metres to several kilometres. They are among the largest structures built by man. The demands on design and on materials are very high. A bridge must be strong enough to support its own weight as well as the weight of the people and vehicles that use it. The structure also must resist various natural occurrences, including earthquakes, strong winds, and changes in temperature. Most bridges have a concrete, steel, or wood framework and an asphalt or concrete roadway on which people and vehicles travel.

Basic Components of a Bridge

The majority of bridges are held up by at least two supports set in the ground. The distance between two adjacent supports is called a span of a bridge. The supports at each end of the bridge are called abutments, and the supports that stand between the abutments are called piers. The total length of the bridge is the distance between the abutments. Most short bridges are supported only by abutments and are known as single-span bridges. Bridges that have one or more piers in addition to the abutments are called multi-span bridges. Most long bridges are multi-span bridges.

In special cases, like in a pontoon bridge no piers or abutments are provided. It is supported by pontoons (flat-bottomed boats) or other portable floats. There are three main components of a typical bridge in general, namely:

- Superstructure
- Substructure
- Foundation.

The above three components have the following basic parts:

Superstructure :

a) The main constituent of any superstructure is the decking, which comprises of a slab, girder, trusses, etc.

b) The handrails, crash barriers, kerbs, etc.

The three basic types of bridge spans are shown below. Any of these spans may be constructed using beams, girders or trusses. Arch bridges are either simple or continuous (hinged). A cantilever bridge may also include a suspended span.

Substructure :

a) Abutments at the extreme ends of the bridge.

b) Piers at intermediate supports in case of multiple span bridges.

c) Bearings and pedestals for the decking.

Foundation :

a) These include foundations for both abutments and piers. Foundations may be of the type open, well, pile, etc.

Apart from the above, river training works and the approaches to a bridge also form a part of a bridge works.

Selection of Bridge Site

It may not be possible always to have a wide choice of sites for a bridge. This is particularly so in case of bridges in urban areas and flyovers. For rivers bridges in rural areas, usually a wider choice may be available.

The characteristics of an ideal site for a bridge across a river are:

- A straight reach of the river;
- Steady river flow without serious whirls and cross currents;
- A narrow channel with firm banks;
- Suitable high banks above high flood level on each side;
- Rock or other hard inerodible strata close to the river bed level;
- Economical approaches which should not be very high or long or liable to flank attacks of the river during floods; the approaches should be free from obstacles such as hills, frequent drainage crossing, sacred places, graveyards, or built up areas or troublesome land acquisition;
- Proximity to a direct alignment of the road to be connected;
- Absence of sharp curves in the approaches;
- Absence of expensive river training works; and
- Avoidance of excessive underwater construction.

For selecting a suitable site for a major bridge, the investigating engineer should make a reconnaissance survey for about one km on the upstream side and one km on the downstream side of the proposed bridge site and should journey along the road for about one km on either side of the road from the bridge site in order to form a general impression of the landscape and to decide on the type of structure best suited to the site. Care should be taken to investigate a number of probable alternative sites and then decide on the site which is likely to serve the needs of the bridge at the least cost. A brief description of the reasons for selection of a particular site should be furnished in the investigation report along with salient details of alternative sites investigated and rejected.

When the river to be crossed is a meandering river, the bridge should be located at a nodal point, i.e. the location where the river regime is constant serving as a fulcrum about which the river channels swing laterally. To the extent possible, it is desirable to align the bridge at right angles to the river, i.e. to provide a square crossing. Sometimes, a skew crossing which is included to the centre line of the river at an angle different from a right angle has to be provided in order to avoid costly land acquisition or sharp curves on the approaches. A skew bridge usually poses more difficulties in design, construction and maintenance. The location of the bridge in relation to the alignment of the approaches may be decided as below : (a) For bridges of total length less than 60 m, the alignment of the approach will govern; (b) For bridges of total length between 60 m and 300 m, both the proper alignment of the approaches and the requirements of a good bridge site should be considered together in ascertaining the appropriate site; and © For major bridge over 300 m length, the requirements of a good bridge site will govern the alignment.

Bridge Classification

The types of bridges vary in total length, the length of their spans, and the weight they can support. Before deciding which kind to build at a particular place, engineers determine the length of the structure and of each span. They also consider the maximum load the bridge will carry and the materials available to construct it. Considering various aspects, bridges may be classified in the following ways :

1. *Function or purpose to be performed :* Based on this bridges can be classified as an aqueduct over a river or a canal, a viaduct which may be road or railway over a valley, a pedestrian or a foot bridge, a highway bridge, a railway bridge, a road cum rail bridge, etc.
2. *Construction material used :* Under this heading bridges can be grouped as timber bridges, masonry bridges, steel bridges, reinforced cement concrete bridges, prestressed bridges, composite bridges, etc.
3. *Alignment of bridge :* Based on this bridges can be classified as a straight or a skew bridge.
4. *Location of bridge floor relative to the superstructure :* In accordance with the relative position of bridge floor as compared to the superstructure, bridges can be classified as deck, through, semi-through and suspended bridge.
5. *According to the type of superstructure :* Depending upon the type of superstructure, bridges can be grouped as slab, girder, truss, suspension, cable stayed, arch, drawbridges, etc.
6. *Nature of superstructure action :* Under its category bridges can be classified as portal frame bridges, truss bridges, balanced cantilever bridges and suspension bridges.
7. *Location of Highest Flood Level :* Particularly for a highway bridge, under this category bridges may be classified as a high level or non - submersible bridge and a submersible bridge depending upon the formation level of the bridge relative to the highest flood level.

8. *Fixed or Movable / Method of clearance for navigation:* In case of navigable channels which don't have permanent and sufficient clear waterway, fixed or movable bridges are used namely high - level fixed, movable bascule, movable swing or transporter bridge.
9. *Type of loadings:* As per the Indian Road Congress (IRC), road bridges and culverts may be classified into Class A, Class B and Class 70 R (tracked or wheeled) according to the loads they are designed to carry.
10. *Span length:* Under this heading bridges are grouped as culverts if span length is less than 8 m, minor bridges for span lengths between 8 m and 30 m, major bridges for span length more than 30 m and long span bridges for span length above 120 m.
11. *Method of connections :* Depending upon the method of connections of various superstructure parts particular in case of steel construction, steel bridges can be classified into pinned connected, welded or riveted bridges.
12. *Degree of Redundancy :* Depending upon the degree of redundancy bridges can be classified as determinate bridges and indeterminate bridges.
13. *Inter span relation :* Under this category bridges can be grouped as simply supported, continuous or cantilever type bridges.
14. *Type of service to be put to :* In accordance with the type of service anticipated and duration of use, bridges can be classified as permanent, temporary, military bridges like pontoon and bailey, etc.
15. *Number of lanes:* In accordance with the number of lanes for which a bridge structure has been constructed, bridges can be classified as single lane, two lanes or three lane bridges, etc.

Choice of Superstructure Type

The basic function of a bridge superstructure is to permit uninterrupted smooth passage of traffic over it and to transmit the loads and to transmit the load and forces to the substructure safely through the bearings. Although it is difficult to stipulate the aesthetic requirements, it should, however, be ensured that the type of superstructure adopted is simple, pleasing to the eye, and blends with the environment. No hard and fast rules can be laid regarding a the economy in cost. The designer should, however, be able to evolve the most economical type of superstructure based on his judgement and experience given the particular conditions prevailing at the particular site at the particular time. The following factors are to be considered while selecting the type of a bridge superstructure.

i. The nature of river or streams
ii. Nature of foundation / founding strata available
iii. The amount and type of traffic
iv. Whether used for navigation purposes
v. Climatic conditions
vi. Hydraulic data
vii. Type of available construction material
viii. Labour available
ix. The available facilities for erections
x. Maintenance provisional
xi. The availability of funds
xii. Time available for construction
xiii. Strategic consideration
xiv. Economic consideration
xv. Aesthetic consideration.

Characteristics and Design Aspects of Various Types of Superstructures

Basic Principle

The superstructure of any bridge must be designed so as to satisfy the geometric and load carrying requirements to be met. The geometric requirements depend on the number and widths of traffic lanes and footpaths for which the deck is to cater to. It also depends on the overall alignment and various horizontal and vertical clearances required above and below the roadway. Once the geometric considerations are decided, the superstructure has to be designed to meet various structural design requirements. These include considerations of strength, stiffness and stability. This first requires estimation of internal forces and moments and displacements which the externally applied forces will cause in the selected scheme and form of the structure and then deciding the final section sizes, and reinforcements (and prestressing, if the structure is prestressed). The former process constitutes 'analysis' and the latter process constitutes 'design'. The analysis is generally done based on elastic behaviour of the structure. The design is then done either on elastic (i.e. working) strength basis or on load-factored (i.e. ultimate) strength basis, also ensuring the serviceability criteria (like limiting the flexural crack-widths, deflections and vibrations). Some structural analysis work is done using empirical formulae. If the structure is very complicated, model analysis is resorted to.

The selection of superstructure type suitable for particular requirements will depend upon the comparative advantages and disadvantages of each type.

There are seven main kinds of bridges: (1) slab / voided slab bridges, (2) girder bridges, (3) arch bridges, (4) cantilever bridges, (5) suspension bridges, (6) cable-stayed bridges, (7) truss bridges and (8) other bridges

Material Used for Constructing Superstructure

The following materials are generally adopted for constructing permanent bridge decks :

a. Reinforced concrete
b. Prestressed concrete
c. Steel
d. Masonry
e. Composite construction using steel and reinforced concrete, or reinforced and prestressed concrete.

Reinforced Concrete Superstructure

The common types of deck sections in this category are :

a. Solid slab
b. Slab and girder (T-beam)
c. Hollow box girder.

The deck arrangements can be simple or continuous spans, or frame, balanced cantilever, free cantilever cast in situ precast or segmental.

Composite and Steel Superstructure

The various types of deck in this category are :

a. Longitudinal and transverse beams, with concrete slab
b. Longitudinal plate girders, transverse beams and concrete slab
c. Longitudinal box girder with concrete slab
d. Steel truss.

The deck arrangement can be simple or continuous spans, or arch or frame. The slab can be non-composite or can be made composite with the longitudinal beams by shear connectors.

Special Superstructures :

a. Orthotropic deck
b. Cable stayed deck
c. Suspension type deck.

Age masonry superstructure are generally of arch type.

The composite type of decks generally involve combination of two dissimilar structural elements. Precast or prefabricated beams are connected to RC slab by special structural elements called shear connectors, acting together as a unit. The beams may be either :

a. Prefabricated structural steel beams/girders or
b. Precast reinforced concrete, or precast prestressed concrete beams.

Comparison of Reinforced Concrete and Prestressed Concrete Superstructures

- In recent years a majority of bridges have been constructed in prestressed concrete. However, reinforced concrete bridges in certain circumstances still, are more economical.
- Prestressed concrete bridges offer a high degree of freedom from cracks, low maintenance, long life; and they are particularly suitable for dynamic loads and vibrations, and offer great resistance to fatigue arising from repetitions of live loading.
- In RC construction, live load normally approximately doubles the stress in steel, while in prestressed concrete this increase is very little. Position of neutral axis in RC decks about which moments of areas from tension and compression side balance remains fixed, but of that in prestressed concrete shifts with variation in applied load.
- The greatest advantage of prestressed concrete for bridges is reduction in weight slender appearance, possibility of precasting in factory conditions either in segments or even in whole spans, possibility of larger spans and freedom from tension and cracking. For long spans, prestressed concrete decks can be nearly comparable in weight with steel decks. It is not possible to quote a minimum prestressed concrete span which can be most economical, it depends on local conditions.

For short bridges the advantage of using smaller amount of concrete because of prestressing may be unimportant since there is always a certain basic sot which cannot be changed by the use of prestressing. In such cases the usual reinforced concrete construction may be used with equal advantages.

Bridge Substructure and Foundation

The ' substructure ' and ' foundation ' are generally known to be as the ' better half ' of a bridge structure. This arises from the fact that in a number of cases the substructure and foundation prove to be more expensive than the superstructure especially in case of bridges over major rivers where foundations are very deep and the function of the superstructure carrying the design loads safely depends on the support given by the foundation.

Even though the foundations are not seen by people in general or by those using the bridges but great expertise goes into the design and construction of these in varied soil types, deep waters, marine clays, turbulent flowing rivers, etc. The foundations sometimes reach depths as much as 100 m below the bed. Foundations are very important part of the bridge and the strata on which they are founded has to be unyielding and capable of sustaining the loads. The bridge piers and foundations are required to sustain various loads due to traffic, wind, temperature, differential settlement, seismic forces, etc. The protection of the substructure and foundation is also of utmost importance to ensure the safety of the bridge against the scour of the bed due to water current.

The bridge superstructure cannot be conceived in isolation but is required to be visualised as an integral part of superstructure, structurally as well as architecturally. They are required not only to be engineered to withstand the multitude of loads but also are required to be carefully architecture to blend or harmonize with the surroundings from aesthetic point of view. Thus an integral design approach is required to achieve an architecturally pleasing structure without compromising on the structural integrity and economy.

Abutments

There are two basic types of abutments; open type or spill through type and closed type or non-spill through type.

Open type abutments are more economical than closed type since they allow the back fill to spill out in between their columns thereby reducing the active earth pressure. But due to this very reason, the backfill may settle down, should it be washed through under heavy rains. In such a case of deep curtain wall should be provided extending downward from the abutment cap, for at least 1.5 m, below the spill line.

Piers or Bents

The term 'Pier' pertains to an intermediate substructure unit between the abutment. A bent is a pier consisting of one or more columns, in the substructure, unit with or without cap.

Pile Bents

Type are economical but are suitable for stream crossing where floating debris is minimal. They may be competitive in placed waters and in dry beds, where soil is poor and footings laid at shallower depths might settle. The preferable minimum width of cap is 75 cm. Pile bents are unsightly for bridges 6 m or more in height and unsuitable if scour is significant.

Solid Piers: They are located in streams where floating debris and rolling bounders accompany the last current. They are preferably for long spans, and may be supported on spread footings, pile foundations or caissons, as appropriate. A pier wall should have a low thickness/height ratio, but should not be less than 30 cm thick. This is the minimum thickness that should be used with two planes of reinforcing steel.

Multicolumn Bents and Multi Columns

They generally support dry land structures. They normally replace pile bents when spread footings are recommended. They may be supported on either spread footings or pile foundations. Columns may be circular rectangular or variable in section to provide architectural effect. The thickness/height ratio of normal height uniform section columns should be in a range of 1/12 to 1/15 for appearance as well as structural reasons. Unnecessarily thick columns can invite high temperature and shrinkage stresses and very slender ones will require buckling analysis.

Single Column Bents and Single Columns

They can sometimes be used to avoid skewed piers. They are adapted to viaducts over city streets where the location of the column may be restricted.

Foundations

Footings are appropriate where scour is low and soil bearing capacity is enough not to create settlement problems, or if rock is available at a shallow depth. In any case, if subsoil water is encountered footings are feasible only if dewatering is possible. Otherwise piles are preferable, which enables transferring the loads directly to lower and firmer substrata, without requiring dewatering. These are costlier than footings, in general. However, in case of deep scour, rolling bounders, floating debris and strong current, the (generally costlier) caissons become necessary. Caissons are relatively massive and generally suit spans not less than 25 m preferably longer.

Choice of Type of Foundation

The choice of foundation for the bridges depends upon the conditions of soil. Cost of substructure increases with the depth of foundation. Similarly, the cost of superstructure per unit length also increases with the span length. Hence the decision on span lengths to depend upon the ratio of the cost of substructure including the foundation versus the cost of superstructure i.e. the girder and the deck and it can be found that it is most economical when the ratio is one. There may be cases where though the cost of foundations may be more than the cost of superstructure practical difficulties in sinking the foundation and limitations may require a reconsideration for the reduction of the total number of foundations. Navigational considerations also will have an influence upon the span. The choice of span has to be suitably made taking all these into consideration.

Foundations can be broadly classified as :

(i) open foundations;

(ii) pile foundations; and

(iii) well foundations.

Open foundations are suitable for small bridges and bridges which re sited in a location where rock or firm founding soil is available at a shallow depth and, at the same time, the scour in the river is not considerable. The next choice, from considerations of economy, are the pile foundation which are again more suitable for being used in cohesive soils and beds subject to minimum scour. Well foundations have been more popularly used in India in view of the expertise available particularly for major and important bridges.

General Design Procedure for Design of Substructure of A Bridge

I. First of all compute the dead load reaction due to the superstructure and superimposed dead load along with the maximum and minimum live load reactions coming on the pier or abutment under consideration. Apart from this also compute the corresponding moments caused due to these loads about the longitudinal as well as the transverse directions or axis of the bridge. For this the critical case is the maximum possible eccentric transfer of these dead and live load reactions along with the braking and temperature forces, etc. to the foundation base.

II. Compute the total vertical reaction at the base of the foundation or at the soft of the pile cap in case of pile foundation under maximum, minimum and no live load conditions, taking into account the upward force due to the effect of buoyancy equal to 50% or 100% weight of water in volume equal to that of the submerged mass. This value is 100% in case of saturated soil and fissured, weathered or weak rock and 50% in case of good rock.

III. Compute the total moment about the bridge's longitudinal axis and the horizontal force in the transverse direction, at various levels and more importantly at the base of the foundation due to possible eccentricity of live and dead loads, force due to water current and afflux, if any.

IV. Compute the total moment about the bridge's transverse axis and the horizontal force in the longitudinal direction, at various levels and more importantly at the base of the foundation due to possible eccentricity of live and dead loads, force due to braking and temperature, water current and afflux, if any. The force due to braking can be calculated depending upon the following cases : (i) In case of simply supported spans on rocker / roller bearings, braking force due to live load on one complete span may be assumed to go to its rocker bearing only such that the corresponding foundation under it will resist force either equal to braking - temperature or half of braking + temperature, the latter case being critical for usually an abutment or a pier supporting unequal spans. (ii) In case of simply supported spans with identical neoprene or elastomeric bearings under each of an individual span, the foundation will resist the sum of half of the value of braking force due to live loads on each of the two spans supported by it and the horizontal temperature force equal to the difference of the product of the shear rating of the neoprene bearings and the deck movements above them for left and right hand side spans respectively. (iii) In case of continuous bridge structures the horizontal force being resisted by the bearings depends upon the arrangement and type of bearings chosen and is case specific.

V. Compute the wind force in the transverse direction acting on the exposed area of the bridge deck and substructure above the lowest bed level. This force is calculated with and without the live loading on the bridge deck. Usually it is sufficient to consider wind on the deck surface area between its soffit and top of the solid parapet or crash barrier, and that of the body of the live load at the given rate as per codal recommendations. Similarly for no live load condition the rate specified by the code is considered. In certain cases the wind force strikes the bridge structure obliquely. In such a case a combination of simultaneously acting wind forces acting in the transverse and longitudinal directions is considered as per codal specifications.

VI. Compute (i) the static equivalent of the horizontal seismic force depending upon the mass of the structure above the earth embedment level which is the maximum scour level in case of hydraulic bridges. The seismic forces are based on the full weight including the submerged portions as long as they are above the embedment level. (ii) The seismic force on a mass can be taken as a certain fraction of its weight, acting through its centroid. This fraction varies depending upon the bridge site being in a severe seismic zone to a non - seismic area as per recommendations of the code. Apart from this a vertical seismic force acting in the upward or downward direction and equal to half of the above mentioned force. This vertical force becomes critical in cases where values of maximum and minimum base pressures are critical. (iii) In case of seismic force acting in the transverse direction, the live load seismic effect is assumed to be acting at 1.2 m above the formation level and that due to the dead load of the deck and superimposed dead load at the deck centroid level. (iv) In case the seismic force is in the direction of the longitudinal axis of the bridge, then the seismic effect due to live load is to be ignored since the braking force is already being considered and any further longitudinal force on the live load will only cause skidding of the wheels.

VII. Estimate the 'active' earth pressure force and moment (at various levels and at the base of foundation) on account of the earthfill retained upto foundation base in case of founding strata being soil or upto rock level in case of rocky founding strata. Passive relief from the front fill is generally to be ignored, but if it is well protected then it may be considered depending upto the actual conditions in situ (but accounting for the negative surcharge angle effect if sloping downwards).

In case of wells where the foundation is usually taken down well into permanent and unexecuted soil (e.g. a caisson taken below the maximum scour level), estimate and take into account the net 'passive less active' earth pressure relief (force and moment) as a result of the soil grip.

VIII. Compute the net vertical load, the net horizontal forces in the longitudinal and transverse directions and the net moments about these two directions, at the base of the foundation (at soffit of pile cap in the case of piles) under each critical load combination. Apart from the pile foundations, calculate the base pressures and the safety factors for stability against overturning and sliding and ensure that the requirements are satisfied. In case one of the parametres is not satisfied, then redesign by taking a new set of dimensions.

IX. In case of piled foundations estimate the maximum and minimum axial loads in the piles by the using rivet-group approach. Ensure that no pile is in tension unless tension-piles are provided. Ensure the structural design of the pile-section for the vertical loads and moments combinations on it. Check for the adequacy of soil resistance around an individual pile, the block failure and group action of the pile group.

X. After satisfying the stability requirements of the foundation, calculate the vertical load and moments at various intermediate critical sections (including at the maximum bending moment section in a caisson) in the foundation structure (this includes whole pier/abutment structure), and then structurally design these sections. Also design various other structural components of the foundations structure, taking into account the respective critical load combinations for these components.

XI. Estimate the vertical loads on each bearing under various load combinations, along with the coexisting longitudinal and transverse horizontal forces and rotations. Design the bearings for these effects. Use may be made of the standard manufacturers' catalogues for certain standard bearings.

Types of Piers and their Design Aspects

Open Piers : They are classified as :

1. Multiple bent or multiple column
2. Pile bent
3. Cylindrical piers
4. Trestle piers.

Multiple Bent : They are often used on ground. The section of wall at the bottom may or may not be used. It is needed in overpass work where traffic runs parallel close to the bent to reduce damage to the columns in case of accident. It is lighter and may be more economical than the solid pier; but it requires more formwork and material.

Pile Bent : The pile bent is used for low piers over unstable ground. They consist of reinforced concrete or steel piles driven into the ground, provided with a capping at their top to support the main girders. They are laterally connected by reinforced concrete or steel braces. The pile is used both for a support by driving to resistance and for a column by projecting above ground. The outside piles are usually battered.

Cylindrical Piers : They consist of mild steel or cast iron cylinders which are filled with concrete. When cylinders are used apart and suitable bracing is provided.

Trestle Piers : They are used for temporary work and for timber work. They are made up of RCC or steel vertical, horizontal and diagonal members. The connections in steel trestle may be either riveted or welded. RCC trestle type pier consists of circular columns with a bent cap at the top. In order to avoid moments being transferred from deck to the columns concrete hinges are introduced between the top of the columns and the bent cap. In case of fly overs and elevated roads where trestle piers are used, it is desirable to provide connecting diaphragms between the column.

The following are some of the special piers :

1. Separate piers
2. Abutment piers
3. Cellular piers
4. Framed piers.

Separate piers : They are sometimes used for high bridges. These may be low piers with steel columns above on concrete piers upto the bridge level. In water, they may be combined with a solid shaft below upto flood water level. They may have the advantage of lower cost than solid shafts.

Abutment Piers : In the case of multiple span arch bridges, every third or fourth pier is designed as an abutment to receive the thrust from either side. Such a pier is thicker in section and is known as abutment pier. The object of the abutment pier is listed below :

i. The arches can be turned in sets, and thus heavy expenditure in centreing is saved

ii. In the event of damage, occurring to any portion of the bridge due to floods, the same will probably not extend further than the next abutment.

Cellular Piers: The cellular piers consists of two concentric RCC Shells connected by radial ribs and horizontal bands at suitable intervals. The intermediate space between the shell is filled with some filler material. This will result in saving of cement concrete which is a costly material. This type of piers usually require difficult shuttering and additional labour in placing reinforcements.

Framed Piers : This type of piers result in reduced effective span length for girders on ether side of the centre line of the pier leading to economy in the cost of superstructure. In this type of design two expansion joints at close intervals to 1 to 2 m are provided which results in uneven surface and maintenance problems. Such type of piers if used in rivers subjected to sudden floods near hills and forests would be subjected to floating debris e.g. floating tree trunk etc.

Design of Piers

The following loads and forces are taken into account :

- Dead weight of the Pier
- End reactions due to dead and live loads of superstructure : These reactions will cause an eccentricity when one span only is loaded. Also some end details are so standardised as to dimension that eccentricity which may be caused by difference in span lengths.
- Impact forces due to floating debris and live load for the top 3 m of the pier only.
- Wind pressure acting on the superstructure, moving loads and the pier.
- Centrifugal forces in the event of a road bridge situated on a curve.
- Longitudinal forces arising either due to tractive effort caused through acceleration of the vehicle or braking effect are resulting from the application of the brakes and due to friction at bearings.
- Water pressure
- Earthquake forces in the event of bridge located in seismic zone.
- Buoyancy effect of submerged part of substructure.

The magnitude of the above forces can be determined as per IRC Standard specifications and code of practice section II. Bridge spanning across a tidal river or an estuary, are subjected to an additional hydrodynamic forces due to wave action. The wave motion is essentially an unsteady fluid flow. The fluid particles are subjected to not only a velocity in the horizontal and vertical directions but also to acceleration in these directions. Thus, the fluid particles velocity and acceleration induce a horizontal drag force and an inertial force. The forces developed in the vertical direction are insignificant because of the piers are normally rigidly fixed at the bottom.

Stability of Bridge Piers

Bridge piers are considered stable if they satisfy the following conditions :

1. All the components parts of the pier must be sufficiently strong to resist bending moments and shearing forces.
2. The pier should be stable against sliding downstream and sliding parallel to the axis of the bridge.
3. The pier must not be turned over about the downstream toe and also in the direction of the axis of the bridge.
4. Under the worst combinations of all loads and forces, no tension should be permitted under the foundations, and the maximum pressure should not exceed the safe bearing capacity of the soil.

Chapter 7

Hydraulic Engineering

Hydraulic engineering as a sub-discipline of civil engineering is concerned with the flow and conveyance of fluids, principally water and sewage. One feature of these systems is the extensive use of gravity as the motive force to cause the movement of the fluids. This area of civil engineering is intimately related to the design of bridges, dams, channels, canals, and levees, and to both sanitary and environmental engineering.

***Figure:** Hydraulic Flood Retention Basin (HFRB)*

Hydraulic engineering is the application of fluid mechanics principles to problems dealing with the collection, storage, control, transport, regulation, measurement, and use of water. Before beginning a hydraulic engineering project, one must figure out how much water is involved. The hydraulic engineer is concerned with the transport of sediment by the river, the interaction of the water with its alluvial boundary, and the occurrence of scour and deposition. "The hydraulic engineer actually develops conceptual designs for the various features which interact with water such a spillways and outlet works for dams, culverts for highways, canals and related structures for irrigation projects, and cooling-water facilities for thermal power plants."

Fundamental Principles

A few examples of the fundamental principles of hydraulic engineering include fluid mechanics, fluid flow, behaviour of real fluids, hydrology, pipelines, open channel hydraulics, mechanics of sediment transport, physical modelling, hydraulic machines, and drainage hydraulics.

Fluid Mechanics

Fundamentals of Hydraulic Engineering defines hydrostatics as the study of fluids at rest. Fluids at rest indicate that there exists a force, known as pressure, that acts upon its surroundings. This pressure, measured in N/m^2, is not constant throughout the body of fluid. Pressure, p, in a given body of fluid, increases with an increase in depth. Where the upward force on a body acts on the base and can be found by equation:

$$p=\rho gy$$

where,

ρ = density of water

g = specific gravity

y = depth of the body of liquid

Rearranging this equation gives you the "pressure head" $p/\rho g = y$. Four basic devices for finding pressure are a piezometre, manometre, differential manometre, Bourdon gauge, as well as an inclined manometre.

As Prasuhn states: On undisturbed submerged bodies, pressure acts along all surfaces of a body in a liquid, causing equal perpendicular forces in the body to act against the pressure of the liquid. This reaction is known as equilibrium. More advanced applications of pressure are that on plane surfaces, curved surfaces, dams, and quadrant gates, just to name a few.

Behaviour of Real Fluids

1. *Real and Ideal fluids :* The main difference between an ideal fluid and a real fluid is that for ideal flow $p^1=p^2$ and for real flow $p^1>p^2$.
2. *Viscous Flow :* A viscous fluid will deform continuously under a shear force, whereas an ideal fluid doesn't deform.
3. *Laminar Flow and Turbulence :* The various effects of disturbance on a viscous flow are stable, transition and unstable.

   Bernouilli's equation:

   $p/\rho g + u^2/2g = p_A/\rho g + u_A{}^2/2g = p_B/\rho g + u_B{}^2/2g$

   where $p_B > p_A$
4. *Boundary Layer :* Assuming a flow is bounded on one side only, and that a rectilinear flow passing over a stationary flat plate which lies parallel to the flow, the flow just upstream of the plate has a uniform velocity. As the flow comes into contact with the plate, the layer of fluid actually 'adheres' to a solid surface. There is then a considerable shearing action between the layer of fluid on the plate surface and the second layer of fluid. The second layer is therefore forced to decelerate (though it is not quite brought to rest), creating a shearing action with the third layer of fluid, and so on. As the fluid passes further along the plate, the zone in which shearing action occurs tends to spread further outwards. This zone is known as the 'boundary layer'. The flow outside the boundary layer is free of shear and viscous-related forces so it is assumed to act like an ideal fluid. "The intermolecular cohesive forces in a fluid are not great enough to hold fluid together. Hence a fluid will flow under the action of the slightest street and flow will continue as long as the stress is present. The flow inside the layer can be either viscous or turbulent, depending on Reynolds number.

Applications

Common topics of design for hydraulic engineers include hydraulic structures such as dams, levees, water distribution networks, water collection networks, sewage collection networks, storm water management, sediment transport, and various other topics related to transportation engineering and geotechnical engineering. Equations developed from the principles of fluid dynamics and fluid mechanics are widely utilised by other engineering disciplines such as mechanical, aeronautical and even traffic engineers.

Related branches include hydrology and rheology while related applications include hydraulic modelling, flood mapping, catchment flood management plans, shoreline management plans, estuarine strategies, coastal protection, and flood alleviation.

History

Earliest uses of hydraulic engineering were to irrigate crops and dates back to the Middle East and Africa. Controlling the movement and supply of water for growing food has been used for many thousands of years. One of the earliest hydraulic machines, the water clock was used in the early 2nd millennium BC. Other early examples of using gravity to move water include the Qanat system in ancient Persia and the very similar Turpan water system in ancient China as well as irrigation canals in Peru. In ancient China, hydraulic engineering was highly developed, and engineers constructed massive canals with levees and dams to channel the flow of water for irrigation, as well as locks to allow ships to pass through. Sunshu Ao is considered the first Chinese hydraulic engineer. Another important Hydraulic Engineer in China, Ximen Bao was credited of starting the practice of large scale canal irrigation during the Warring States Period (481 BC-221 BC), even today hydraulic engineers remain a respectable position in China. Before becoming President, Hu Jintao was a hydraulic engineer and holds an engineering degree from Tsinghua University

Eupalinos of Megara, was an ancient Greek engineer who built the Tunnel of Eupalinos on Samos Island in the 6th century BC, an important feat of both civil and hydraulic engineering. The civil engineering aspect of this tunnel was the fact that it was dug from both ends which required the diggers to maintain an accurate path so that the two tunnels met and that the entire effort maintained a sufficient slope to allow the water to flow.

Hydraulic engineering was highly developed in Europe under the aegis of the Roman Empire where it was especially applied to the construction and maintenance of aqueducts to supply water to and remove sewage from their cities. In addition to supplying the needs of their citizens they used hydraulic mining methods to prospect and extract alluvial gold deposits in a technique known as hushing, and applied the methods to other ores such as those of tin and lead.

Further advances in hydraulic engineering occurred in the Muslim world between the 8th to 16th centuries, during what is known as the Islamic Golden Age. Of particular importance was the 'water management technological complex' which was central to the Islamic

Green Revolution and, by extension, a precondition for the emergence of modern technology. The various components of this 'toolkit' were developed in different parts of the Afro-Eurasian landmass, both within and beyond the Islamic world. However, it was in the medieval Islamic lands where the technological complex was assembled and standardised, and subsequently diffused to the rest of the Old World. Under the rule of a single Islamic Caliphate, different regional hydraulic technologies were assembled into "an identifiable water management technological complex that was to have a global impact." The various components of this complex included canals, dams, the *qanat* system from Persia, regional water-lifting devices such as the *noria, shaduf* and screwpump from Egypt, and the windmill from Islamic Afghanistan. Other original Islamic developments included the *saqiya* with a flywheel effect from Islamic Spain, the reciprocating suction pump and crankshaft-connecting rod mechanism from Iraq, the geared and hydropowered water supply system from Syria, and the water purification methods of Islamic chemists.

Modern Times

In many respects the fundamentals of hydraulic engineering haven't changed since ancient times. Liquids are still moved for the most part by gravity through systems of canals and aqueducts, though the supply reservoirs may now be filled using pumps. The need for water has steadily increased from ancient times and the role of the hydraulic engineer is a critical one in supplying it. For example, without the efforts of people like William Mulholland the Los Angeles area would not have been able to grow as it has because it simply doesn't have enough local water to support its population. The same is true for many of our world's largest cities. In much the same way, the central valley of California could not have become such an important agricultural region without effective water management and distribution for irrigation.

Leonard Da Vinci (1452-1519) performed experiments, investigated and speculated on waves and jets, eddies and streamlining. Isaac Newton (1642-1727) by formulating the laws of motion and his law of viscosity, in addition to developing the calculus, paved the way for many great developments in fluid mechanics. Using Newton's laws of motion, numerous 18th century mathematicians solved many frictionless (zero-viscosity) flow problems. However, most flows are dominated by viscous effects, so engineers of the 17th and 18th centuries found the inviscid flow solutions unsuitable, and by experimentation they developed empirical equations, thus establishing the science of hydraulics. Late in the 19th century, the importance of dimensionless numbers and their relationship to turbulence was recognised, and dimensional analysis was born. In 1904 Ludwig Prandtl published a key paper, proposing that the flow fields of low-viscosity fluids be divided into two zones, namely a thin, viscosity-dominated boundary layer near solid surfaces, and an effectively inviscid outer zone away from the boundaries. This concept explained many former paradoxes, and enabled subsequent engineers to analyse far more complex flows. However, we still have no complete theory for the nature of turbulence, and so modern fluid mechanics continues to be combination of experimental results and theory. In a somewhat parallel way to what happened in California the creation of the Tennessee Valley Authority (TVA) brought work and prosperity to the South by building dams to generate cheap electricity and control flooding in the region, making rivers navigable and generally modernising life the region.

The modern hydraulic engineer uses the same kinds of computer-aided design (CAD) tools as many of the other engineering disciplines while also making use of technologies like computational fluid dynamics to perform the calculations to accurately predict flow characteristics, GPS mapping to assist in locating the best paths for installing a system and laser-based surveying tools to aid in the actual construction of a system.

Materials Science

Materials science is an interdisciplinary field applying the properties of matter to various areas of science and engineering. This scientific field investigates the relationship between the structure of materials at atomic or molecular scales and their macroscopic properties. It incorporates elements of applied physics and chemistry. With significant media attention focused on nanoscience and nanotechnology in recent years, materials science has been propelled to the forefront at many universities. It is also an important part of forensic engineering and failure analysis. Materials science also deals with *fundamental properties* and *characteristics* of materials.

History

The material of choice of a given era is often a defining point. Phrases such as Stone Age, Bronze Age, and Steel Age are good examples. Originally deriving from the manufacture of ceramics and its putative derivative metallurgy, materials science is one of the oldest forms of engineering and applied science. Modern materials science evolved directly from metallurgy, which itself evolved from mining and (likely) ceramics and the use of fire. A major breakthrough in the understanding of materials occurred in the late 19th century, when the American scientist Josiah Willard Gibbs demonstrated that the thermodynamic properties related to atomic structure in various phases are related to the physical properties of a material.

Important elements of modern materials science are a product of the space race: the understanding and engineering of the metallic alloys, and silica and carbon materials, used in the construction of space vehicles enabling the exploration of space. Materials science has driven, and been driven by, the development of revolutionary technologies such as plastics, semiconductors, and biomaterials. Before the 1960's (and in some cases decades after), many *materials science* departments were named *metallurgy* departments, from a 19th and early 20th century emphasis on metals. The field has since broadened to include every class of materials, including: ceramics, polymers, semiconductors, magnetic materials, medical implant materials and biological materials (materiomics).

Fundamentals

The basis of materials science involves relating the desired properties and relative performance of a material in a certain application to the structure of the atoms and phases in that material through characterisation. The major determinants of the structure of a material and thus of its properties are its constituent chemical elements and the way in which it has been processed into its final form. These characteristics, taken together and related through the laws of thermodynamics, govern a material's microstructure, and thus its properties. The manufacture of a perfect crystal of a material is currently physically impossible. Instead materials scientists manipulate the defects in crystalline materials such as precipitates, grain boundaries (Hall–Petch relationship), interstitial atoms, vacancies or substitutional atoms, to create materials with the desired properties.

Not all materials have a regular crystal structure. Polymers display varying degrees of crystallinity, and many are completely non-crystalline. Glasses, some ceramics, and many natural materials are amorphous, not possessing any long-range order in their atomic arrangements. The study of polymers combines elements of chemical and statistical thermodynamics to give thermodynamic, as well as mechanical, descriptions of physical properties.

In addition to industrial interest, materials science has gradually developed into a field which provides tests for condensed matter or solid state theories. New physics emerge because of the diverse new material properties which need to be explained.

Classes of Materials

Materials science encompasses various classes of materials, each of which may constitute a separate field. There are several ways to classify materials. For instance by the type of bonding between the atoms. The traditional groups are ceramics, metals and polymers based on atomic structure and chemical composition. New materials has resulted in more classes. One way of classifying materials is:

- Biomaterials
- Ceramics
- Composite materials
- Metals
- Polymers
- Semiconductors.

Materials in Industry

Radical materials advances can drive the creation of new products or even new industries, but stable industries also employ materials scientists to make incremental improvements and troubleshoot issues with currently used materials. Industrial applications of materials science include materials design, cost-benefit trade-offs in industrial production of materials, processing techniques (casting, rolling, welding, ion implantation, crystal growth, thin-film deposition, sintering, glass-blowing, etc.), and analytical techniques (characterisation techniques such as electron microscope, x-ray diffraction, calorimetry, nuclear microscope (HEFIB), Rutherford backscattering, neutron diffraction, small-angle X-ray scattering (SAXS), etc.).

Besides material characterisation, the material scientist/engineer also deals with the extraction of materials and their conversion into useful forms. Thus ingot casting, foundry techniques, blast furnace extraction, and electrolytic extraction are all part of the required knowledge of a metallurgist/engineer. Often the presence, absence or variation of minute quantities of secondary elements and compounds in a bulk material will have a great impact on the final properties of the materials produced, for instance, steels are classified based on 1/10 and 1/100 weight percentages of the carbon and other alloying elements they contain. Thus, the extraction and purification techniques employed in the extraction of iron in the blast furnace will have an impact of the quality of steel that may be produced.

The overlap between physics and materials science has led to the offshoot field of *materials physics*, which is concerned with the physical properties of materials. The approach is generally more macroscopic and applied than in condensed matter physics. See important publications in materials physics for more details on this field of study.

Ceramics and Glasses

Another application of the material sciences is the structures of glass and ceramics, typically associated with the most brittle materials. Bonding in ceramics and glasses are using covalent and ionic-covalent types with SiO_2 (silica or sand) as a fundamental building block. Ceramics are as soft as clay and as hard as stone and concrete. Usually, they are crystalline in form. Most glasses contain a metal oxide fused with silica. At high temperatures used to prepare glass, the material is a viscous liquid. The structure of glass forms into an amorphous state upon cooling. Windowpanes and eyeglasses are important examples. Fibres of glass are also available. Diamond and carbon in its graphite form are considered to be ceramics.

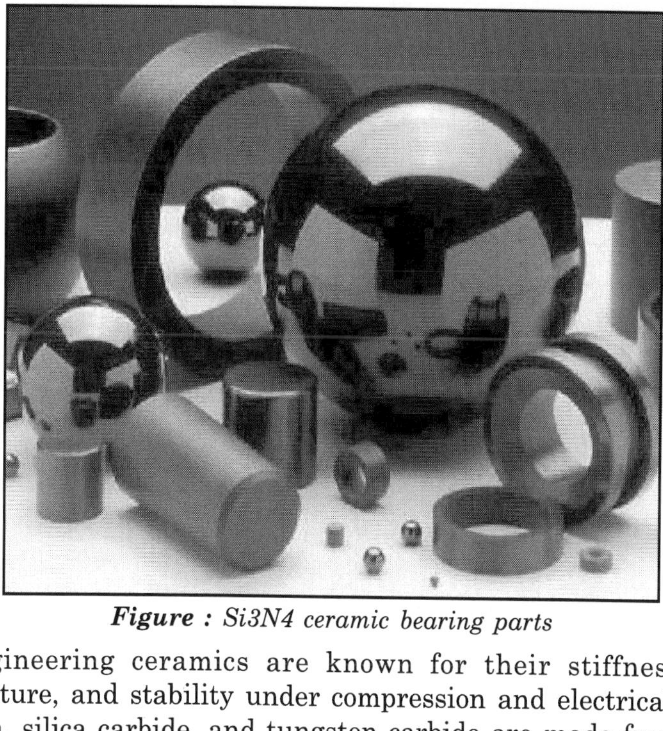

Figure : *Si3N4 ceramic bearing parts*

Engineering ceramics are known for their stiffness, high temperature, and stability under compression and electrical stress. Alumina, silica carbide, and tungsten carbide are made from a fine powder of their constituents in a process of sintering with a binder.

Hot pressing provides higher density material. Chemical vapour deposition can place a film of a ceramic on another material. Cermets are ceramic particles containing some metals. The wear resistance of tools is derived from cemented carbides with the metal phase of cobalt and nickel typically added to modify properties.

Composite Materials

Another application of material science in industry is the making of composite materials. Composite materials are structured materials composed of two or more macroscopic phases. Applications range from structural elements such as steel-reinforced concrete, to the thermally insulative tiles which play a key and integral role in NASA's Space Shuttle thermal protection system which is used to protect the surface of the shuttle from the heat of re-entry into the Earth's atmosphere. One example is reinforced Carbon-Carbon (RCC), The light gray material which withstands re-entry temperatures up to 1510 °C (2750 °F) and protects the Space Shuttle's wing leading edges and nose cap. RCC is a laminated composite material made from graphite rayon cloth and impregnated with a phenolic resin. After curing at high temperature in an autoclave, the laminate is pyrolised to convert the resin to carbon, impregnated with furfural alcohol in a vacuum chamber, and cured/pyrolised to convert the furfural alcohol to carbon. In order to provide oxidation resistance for reuse capability, the outer layers of the RCC are converted to silicon carbide.

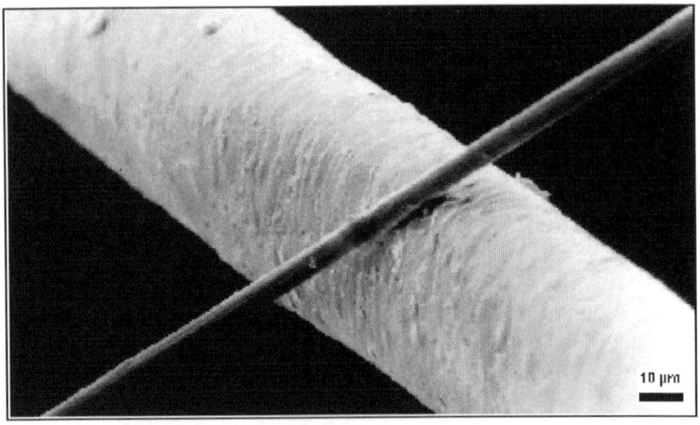

***Figure** : A 6 μm diameter carbon filament (running from bottom left to top right) compared to a human hair.*

Other examples can be seen in the "plastic" casings of television sets, cell-phones and so on. These plastic casings are usually a composite material made up of a thermoplastic matrix such as acrylonitrile butadiene styrene (ABS) in which calcium carbonate chalk, talc, glass fibres or carbon fibres have been added for added strength, bulk, or electrostatic dispersion. These additions may be referred to as reinforcing fibres, or dispersants, depending on their purpose.

Polymers

Polymers are also an important part of materials science. Polymers are the raw materials (the resins) used to make what we commonly call plastics. Plastics are really the final product, created after one or more polymers or additives have been added to a resin during processing, which is then shaped into a final form. Polymers which have been around, and which are in current widespread use, include polyethylene, polypropylene, PVC, polystyrene, nylons, polyesters, acrylics, polyurethanes, and polycarbonates. Plastics are generally classified as "commodity", "speciality" and "engineering" plastics. PVC (polyvinyl-chloride) is widely used, inexpensive, and annual production quantities are large. It lends itself to an incredible array of applications, from artificial leather to electrical insulation and cabling, packaging and containers. Its fabrication and processing are simple and well-established. The versatility of PVC is due to the wide range of plasticisers and other additives that it accepts. The term "additives" in polymer science refers to the chemicals and compounds added to the polymer base to modify its material properties.

Polycarbonate would be normally considered an engineering plastic (other examples include PEEK, ABS). Engineering plastics are valued for their superior strengths and other special material properties. They are usually not used for disposable applications, unlike commodity plastics. Speciality plastics are materials with unique characteristics, such as ultra-high strength, electrical conductivity, electro-fluorescence, high thermal stability, etc. The dividing lines between the various types of plastics is not based on material but rather on their properties and applications. For instance, polyethylene (PE) is a cheap, low friction polymer commonly used to make disposable shopping bags and trash bags, and is considered a commodity plastic, whereas medium-density polyethylene (MDPE) is used for underground gas and water pipes, and another variety called Ultra-high Molecular Weight Polyethylene UHMWPE is an engineering plastic which is used extensively as the glide rails for industrial equipment and the low-friction socket in implanted hip joints.

Metal Alloys

The study of metal alloys is a significant part of materials science. Of all the metallic alloys in use today, the alloys of iron (steel, stainless steel, cast iron, tool steel, alloy steels) make up the largest proportion both by quantity and commercial value. Iron alloyed with various proportions of carbon gives low, mid and high carbon steels. An iron carbon alloy is only considered steel if the carbon level is between 0.01% and 2.00%. For the steels, the hardness and tensile strength of the steel is related to the amount of carbon present, with increasing carbon levels also leading to lower ductility and toughness. Heat treatment processes such as quenching and tempering can significantly change these properties however. Cast Iron is defined as an iron–carbon alloy with more than 2.00% but less than 6.67% carbon. Stainless steel is defined as a regular steel alloy with greater than 10% by weight alloying content of Chromium. Nickel and Molybdenum are typically also found in stainless steels.

Other significant metallic alloys are those of aluminium, titanium, copper and magnesium. Copper alloys have been known for a long time (since the Bronze Age), while the alloys of the other three metals have been relatively recently developed. Due to the chemical reactivity of these metals, the electrolytic extraction processes required were only developed relatively recently. The alloys of aluminium, titanium and magnesium are also known and valued for their high strength-to-weight ratios and, in the case of magnesium, their ability to provide electromagnetic shielding. These materials are ideal for situations where high strength-to-weight ratios are more important than bulk cost, such as in the aerospace industry and certain automotive engineering applications.

Structural Engineering

Structural engineering is a field of engineering dealing with the analysis and design of structures that support or resist loads. Structural engineering is usually considered a speciality within civil engineering, but it can also be studied in its own right. Structural engineers are most commonly involved in the design of buildings and large nonbuilding structures but they can also be involved in the design of machinery, medical equipment, vehicles or any item where structural integrity affects the item's function or safety. Structural engineers must ensure their designs satisfy given design criteria, predicated on safety (e.g. structures must not collapse without due warning) or serviceability and performance (e.g. building sway must not cause discomfort to the occupants). Buildings are made to endure massive loads as well as changing climate and natural disasters.

Structural engineering theory is based upon physical laws and empirical knowledge of the structural performance of different landscapes and materials. Structural engineering design utilises a relatively small number of basic structural elements to build up structural systems that can be very complex. Structural engineers are responsible for making creative and efficient use of funds, structural elements and materials to achieve these goals.

A Structural Engineer

Structural engineers are responsible for engineering design and analysis. Entry-level structural engineers may design the individual structural elements of a structure, for example the beams, columns, and floors of a building. More experienced engineers would be responsible for the structural design and integrity of an entire system, such as a building. Structural engineers often specialize in particular fields, such as bridge engineering, building engineering, pipeline engineering, industrial structures, or special mechanical structures such as vehicles or aircraft.

Structural engineering has existed since humans first started to construct their own structures. It became a more defined and formalised profession with the emergence of the architecture profession as distinct from the engineering profession during the industrial revolution in the late 19th century. Until then, the architect and the structural engineer were usually one and the same - the master builder. Only with the development of specialised knowledge of structural theories that emerged during the 19th and early 20th centuries did the professional structural engineer come into existence.

The role of a structural engineer today involves a significant understanding of both static and dynamic loading, and the structures that are available to resist them. The complexity of modern structures often requires a great deal of creativity from the engineer in order to ensure the structures support and resist the loads they are subjected to. A structural engineer will typically have a four or five year undergraduate degree, followed by a minimum of three years of professional practice before being considered fully qualified.

Structural engineers are licensed or accredited by different learned societies and regulatory bodies around the world (for example, the Institution of Structural Engineers in the UK). Depending on the degree course they have studied and/or the jurisdiction they are seeking licensure in, they may be accredited (or licensed) as just structural engineers, or as civil engineers, or as both civil and structural engineers.

Earthquake Engineering Structures

Earthquake engineering structures are those engineered to withstand various types of hazardous earthquake exposures at the sites of their particular location. Earthquake engineering is treating its subject structures like defensive fortifications in military engineering but for the warfare on earthquakes. Both earthquake and military general design principles are similar: be ready to slow down or mitigate the advance of a possible attacker.

The main objectives of earthquake engineering are:

- Understand interaction of structures with the shaky ground.
- Foresee the consequences of possible earthquakes.
- Design and construct the structures to perform while being exposed to an earthquake.

Earthquake engineering or earthquake-proof structure does not, necessarily, mean *extremely strong* and *expensive* one like El Castillo pyramid at Chichen Itza shown above. In fact, many structures considered *strong* may in fact be actually *stiff,* which may result in poor seismic performance. Now, the most *powerful* and *budgetary* tool of the earthquake engineering is base isolation which pertains to the passive structural vibration control technologies.

Civil Engineering Structures

Civil structural engineering includes all structural engineering related to the built environment. It includes:

- Bridges
- Dams
- Earthworks
- Foundations
- Offshore structures
- Pipelines
- Power stations
- Railways
- Retaining structures and walls
- Roads
- Tunnels
- Waterways
- Water and wastewater infrastructure.

The structural engineer is the lead designer on these structures, and often the sole designer. In the design of structures such as these, structural safety is of paramount importance (in the UK, designs for dams, nuclear power stations and bridges must be signed off by a chartered engineer). Civil engineering structures are often subjected to very extreme forces, such as large variations in temperature, dynamic loads such as waves or traffic, or high pressures from water or compressed gases. They are also often constructed in corrosive environments, such as at sea, in industrial facilities or below ground.

Mechanical Structures

Principals of structural engineering are applied to variety of mechanical (moveable) structures. The design of static structures assumes they always have the same geometry (in fact, so-called static structures can move significantly, and structural engineering design must take this into account where necessary), but the design of moveable or moving structures must account for fatigue, variation in the method in which load is resisted and significant deflections of structures. The forces which parts of a machine are subjected to can vary significantly, and can do so at a great rate. The forces which a boat or aircraft are subjected to vary enormously and will do so thousands of times over the structure's lifetime. The structural design must ensure that such structures are able to endure such loading for their entire design life without failing.

These works can require mechanical structural engineering:

- Airframes and fuselages
- Boilers and pressure vessels
- Coachworks and carriages
- Cranes
- Elevators
- Escalators
- Marine vessels and hulls.

Structural Elements

Any structure is essentially made up of only a small number of different types of elements:

- Columns
- Beams
- Plates
- Arches

- Shells
- Catenaries

Columns

Columns are elements that carry only axial force - either tension or compression - or both axial force and bending (which is technically called a beam-column but practically, just a column). The design of a column must check the axial capacity of the element, and the buckling capacity. The buckling capacity is the capacity of the element to withstand the propensity to buckle. Its capacity depends upon its geometry, material, and the effective length of the column, which depends upon the restraint conditions at the top and bottom of the column. The effective length is $K * l$ where l is the real length of the column. The capacity of a column to carry axial load depends on the degree of bending it is subjected to, and vice versa. This is represented on an interaction chart and is a complex nonlinear relationship.

Beams

A beam may be defined as an element in which one dimension is much greater than the other two and the applied loads are usually normal to the main axis of the element. Beams and columns are called line elements and are often represented by simple lines in structural modelling.

- cantilevered (supported at one end only with a fixed connection)
- simply supported (supported vertically at each end; horizontally on only one to withstand friction, and able to rotate at the supports)
- continuous (supported by three or more supports)
- a combination of the above (ex. supported at one end and in the middle)

Beams are elements which carry pure bending only. Bending causes one part of the section of a beam (divided along its length) to go into compression and the other part into tension. The compression part must be designed to resist buckling and crushing, while the tension part must be able to adequately resist the tension.

Struts and Ties

A truss is a structure comprising two types of structural elements; compression members and tension members (i.e. struts and ties). Most trusses use gusset plates to connect intersecting elements. Gusset plates are relatively flexible and minimize bending moments at the connections, thus allowing the truss members to carry primarily tension or compression. Trusses are usually utilised in span large distances, where it would be uneconomical to use solid beams.

Plates

Plates carry bending in two directions. A concrete flat slab is an example of a plate. Plates are understood by using continuum mechanics, but due to the complexity involved they are most often designed using a codified empirical approach, or computer analysis.

They can also be designed with yield line theory, where an assumed collapse mechanism is analysed to give an upper bound on the collapse load. This is rarely used in practice.

Shells

Shells derive their strength from their form, and carry forces in compression in two directions. A dome is an example of a shell. They can be designed by making a hanging chain model, which will act as a catenary in pure tension, and inverting the form to achieve pure compression.

Arches

Arches carry forces in compression in one direction only, which is why it is appropriate to build arches out of masonry. They are designed by ensuring that the line of thrust of the force remains within the depth of the arch. Its mainly used to increased the bountifulness of any structure.

Catenaries

Catenaries derive their strength from their form, and carry transverse forces in pure tension by deflecting (just as a tightrope will sag when someone walks on it). They are almost always cable or fabric structures. A fabric structure acts as a catenary in two directions.

Structural Engineering Theory

Structural engineering depends upon a detailed knowledge of applied mechanics, materials science and applied mathematics to understand and predict how structures support and resist self-weight and imposed loads.

To apply the knowledge successfully a structural engineer generally requires detailed knowledge of relevant empirical and theoretical design codes, the techniques of structural analysis, as well as some knowledge of the corrosion resistance of the materials and structures, especially when those structures are exposed to the external environment.

Since the 1990's, specialist software has become available to aid in the design of structures, with the functionality to assist in the drawing, analysing and designing of structures with maximum precision; examples include Auto CAD, StaadPro, ETABS, Prokon etc. Such software may also take into consideration environmental loads, such as from earthquakes and winds.

Materials

Structural engineering depends on the knowledge of materials and their properties, in order to understand how different materials support and resist loads.

Common structural materials are:

- Iron:
  - Wrought iron
  - Cast iron
  - Steel
  - Stainless steel
- Concrete:
  - Reinforced concrete
  - Prestressed concrete
- Aluminium
- Composite materials
- Alloy
- Masonry
- Timber
- Other structural materials:
  - Adobe
  - Bamboo
  - Carbon fibre
  - Fibre reinforced plastic
  - Mudbrick
  - Roofing materials.

Chapter 8

Moment of Inertia

In classical mechanics, moment of inertia, also called mass moment of inertia, rotational inertia, polar moment of inertia of mass, or the angular mass, (SI units kg·m^2) is a measure of an object's resistance to changes to its rotation. It is the inertia of a rotating body with respect to its rotation. The moment of inertia plays much the same role in rotational dynamics as mass does in linear dynamics, describing the relationship between angular momentum and angular velocity, torque and angular acceleration, and several other quantities. The symbol I and sometimes J are usually used to refer to the moment of inertia or polar moment of inertia. While a simple scalar treatment of the moment of inertia suffices for many situations, a more advanced tensor treatment allows the analysis of such complicated systems as spinning tops and gyroscopic motion.

The concept was introduced by Leonhard Euler in his book *Theoria motus corporum solidorum seu rigidorum* in 1765. In this book, he discussed the moment of inertia and many related concepts, such as the principal axis of inertia.

The moment of inertia of an object about a given axis describes how difficult it is to change its angular motion about that axis. Therefore, it encompasses not just how much mass the object has overall, but how far each bit of mass is from the axis. The farther out the object's mass is, the more rotational inertia the object has, and the more torque (force* distance from axis of rotation) is required to change its rotation rate. For example, consider two hoops, A and B, made of the same material and of equal mass. Hoop A is larger in diametre but thinner than B. It requires more effort to accelerate hoop A (change its angular velocity) because its mass is distributed farther from its axis of rotation: mass that is farther out from that axis must, for a given angular velocity, move more quickly than mass closer in. So in this case, hoop A has a larger moment of inertia than hoop B.

The moment of inertia of an object can change if its shape changes. Figure skaters who begin a spin with arms outstretched provide a striking example. By pulling in their arms, they reduce their moment of inertia, causing them to spin faster (by the conservation of angular momentum). The moment of inertia has two forms, a scalar form, I, (used when the axis of rotation is specified) and a more general tensor form that does not require the axis of rotation to be specified. The scalar moment of inertia, I, (often called simply the "moment of inertia") allows a succinct analysis of many simple problems in rotational dynamics, such as objects rolling down inclines and the behaviour of pulleys. For instance, while a block of any shape will slide down a frictionless decline at the same rate, rolling objects may descend at different rates, depending on their moments of inertia. A hoop will descend more slowly than a solid disk of equal mass and radius because more of its mass is located far from the axis of rotation. However, for (more complicated) problems in which the axis of rotation can change, the scalar treatment is inadequate, and the tensor treatment must be used (although shortcuts are possible in special situations). Examples requiring such a treatment include gyroscopes, tops, and even satellites, all objects whose alignment can change.

The moment of inertia is also called the mass moment of inertia (especially by mechanical engineers) to avoid confusion with the second moment of area, which is sometimes called the area moment of inertia (especially by structural engineers). The easiest way to differentiate these quantities is through their units ($kg{\cdot}m^2$ as opposed to m^4). In addition, moment of inertia should not be confused with polar moment of inertia (more specifically, polar moment of inertia of area), which is a measure of an object's ability to resist torsion (twisting) only, although, mathematically, they are similar: if the solid for which the moment of inertia is being calculated has uniform thickness in the direction of the rotating axis, and also has uniform mass density, the difference between the two types of moments of inertia is a factor of mass per unit area.

Scalar Moment of Inertia for Single Body

Consider a massless rigid rod of length l with a point mass m at one end and rotating about the other end. Suppose the rod rotates at a constant rate so that the mass moves at speed v. Then the kinetic energy T of the mass is:

$$T = \frac{1}{2}mv^2$$

Using $v = l$ù, where ù is the angular velocity, one obtains:

$$T = \frac{1}{2}m(\omega l)^2$$

which can be rearranged to give:

$$T = \frac{1}{2}ml^2\omega^2$$

This equation resembles the original expression for the kinetic energy, but in the place of the linear velocity v is the angular velocity ù, and instead of the mass m is ml^2. The quantity ml^2 can therefore be seen as an analogue of mass for rotational motion; in other words, it is a measure of rotational inertia.

Scalar Moment of Inertia for Many Bodies

Consider a rigid body rotating with angular velocity *ù* around a certain axis. The body consists of N point masses m_i whose distances to the axis of rotation are denoted r_i. Each point mass will have the speed v_i = *ù*r_i, so that the total kinetic energy T of the body can be calculated as

$$T = \sum_{i=1}^{N} \frac{1}{2} m_i v_i^2 = \sum_{i=1}^{N} \frac{1}{2} m_i (\omega r_i)^2 = \frac{1}{2}\omega^2 \left(\sum_{i=1}^{N} m_i r_i^2\right).$$

In this expression the quantity in parentheses is called the moment of inertia of the body (with respect to the specified axis of rotation). It is a purely geometric characteristic of the object, as it depends only on its shape and the position of the rotation axis. The moment of inertia is usually denoted with the capital letter I:

$$I = \sum_{i=1}^{N} m_i r_i^2 .$$

It is worth emphasizing that r_i here is the distance from a point to the *axis of rotation*, not to the origin. As such, the moment of inertia will be different when considering rotations about different axes.

Similarly, the moment of inertia of a continuous solid body rotating about a known axis can be calculated by replacing the summation with the integral:

$$I = \int_V \rho(\mathrm{r}) d(\mathrm{r})^2 \, \mathrm{d}V(\mathrm{r}),$$

where r is the radius vector of a point within the body, $\rho(\mathrm{r})$ is the mass density at point r, and $d(\mathrm{r})$ is the distance from point r to the axis of rotation. The integration is evaluated over the volume V of the body.

Moment of Inertia about a Point

In discussions of the virial theorem, a different definition of moment of inertia is introduced, often qualified as moment of inertia about a point (as opposed to an axis). The defining equations look the same as the ones above, but r_i and $d(\mathrm{r})$ in these cases are understood to be the distance to the origin.

Moment of Inertia Theorems

Calculations for the Moment of Inertia (MOI) of a body are not easy in general. The process can be simplified in the following ways:

- Choosing axes to take advantage of geometric symmetries
- Physical homogeneity (i.e. uniform mass distribution) making the density function $\rho(\mathrm{r})$ become constant (elementary calculations, or generally an approximation)
- Use of the following theorems, see below for details.

Properties

Superposition

The moment of inertia of the body is additive. That is, if a body can be decomposed (either physically or conceptually) into several constituent parts, then the moment of inertia of the whole body about a given axis is equal to the sum of moments of inertia of each part around the same axis.

Basing just on the dimensional analysis, the moment of inertia must take the form $I = c \cdot M \cdot L^2$, where M is the mass, L is the "size" of the body in the direction perpendicular to the axis of rotation, and c is a dimensionless *inertial constant*. Additionally, the length $R=\sqrt{I/M}$ is called the *radius of gyration* of the body.

Perpendicular Axes

If I_x, I_y, I_z are moments of inertia around three perpendicular axes passing through the body's centre of mass, then each of them cannot be greater than the sum of two others: for example I_z d" $I_x + I_y$. Here the equality holds only if the body is flat and located in the *Oxy* coordinate plane.

Parallel Axes

If the object's moment of inertia I_{COM} around a certain axis passing through the centre of mass is known, then the parallel axis theorem or Huygens Steiner theorem provides a convenient formula to compute the moment of inertia I_d of the same body around a different axis, which is parallel to the original and located at a distance *d* from it. The formula is only suitable when the initial and final axes are parallel. In order to compute the moment of inertia about an arbitrary axis, one has to use the object's moment of inertia tensor.

Energy, Angular Momentum, Torque

The rotational kinetic energy of a rigid body with angular velocity *ù* (in radians per second) is expressed in terms of the object's moment of inertia:

$$E_r = \frac{1}{2} I \omega^2.$$

This formula is similar to the translational kinetic energy E_t = ½mv^2. Thus, the moment of inertia *I* plays the role of mass in the rotational dynamics. One key difference between the mass and the (scalar) moment of inertia is that the latter depends on the axis rotation and so is not truly invariant. The invariant characteristic of the body in rotational motion is the tensor of moment of inertia I, defined later.

The angular momentum of the body rotating around one of its principal axis is also proportional to the moment of inertia:

$$L = I\omega.$$

This expression is parallel to the formula for translational momentum $p = mv$, where the moment of inertia *I* plays the role of mass *m*, and the angular velocity *ù* stands for the velocity *v*. The scalar formula is valid for rotations around one of the principal axes of the body or for rotation about any fixed axis. The equivalent formula involving the tensor moment of inertia is always correct and must be used in cases of free rotation about a non-principal axis.

Also when the body rotates around one of its principal axes, and the direction of the axis of rotation remains constant, one can relate the torque on an object and its angular acceleration in a similar equation:

$$\tau = I\alpha,$$

where τ is the torque and α is the angular acceleration.

Examples

Diatomic molecule, with atoms m_1 and m_2 at a distance d from each other, rotating around the axis which passes through the molecule's centre of mass and is perpendicular to the direction of the molecule.The easiest way to calculate this molecule's moment of inertia is to use the parallel axis theorem. If we consider rotation around the axis passing through the atom m_1, then the moment of inertia will be $I_1 = m_1 \cdot 0 + m_2 \cdot d^2 = m_2 d^2$. On the other hand, by the parallel axis theorem this moment is equal to $I_1 = I + /$ m2 + $m_2) \cdot a^2$, where I is the moment of inertia around the axis passing through the centre of mass, and a is the distance between the centre of mass and the first atom. By the centre of mass formula, this distance is equal to $a = m_2 d$ / / m2 + m_2). Thus,

$$I = I_1 - (m_1 + m_2)a^2 = m_2 d^2 - \frac{m_2^2 d^2}{m_1 + m_2} = \frac{m_1 m_2}{m_1 + m_2} d^2.$$

Thin rod of mass m and length –!, rotating around the axis which passes through its centre and is perpendicular to the rod.

Let Oz be the axis of rotation, and Ox the axis along the rod. If ρ is the density, and s the cross-section of the rod (so that $m = \rho \ell s$), then the volume element for the integral formula will be equal to dV = $s \cdot$dx, where x changes from $-½\ell$ to $½\ell$. The moment of inertia can be found by computing the integral:

$$I = \int_{-\ell/2}^{\ell/2} \rho x^2 s \mathrm{d}x = \rho s \frac{x^3}{3} \Big|_{-\ell/2}^{\ell/2} = \frac{m}{s\ell} \cdot s \cdot 2 \frac{\ell^3/8}{3} = \frac{1}{12} m\ell^2.$$

Solid ball of mass m and radius R, rotating around an axis which passes through the centre.

Suppose Oz is the axis of rotation. The distance from point r = (x, y, z) towards the axis Oz is equal to $d(\mathrm{r})^2 = x^2 + y^2$. Thus, in order to compute the moment of inertia I_z, we need to evaluate the integral -"($x^2 + y^2$) dV. The calculation considerably simplifies if we notice that by symmetry of the problem, the moments of inertia around all axes are equal: $I_x = I_y = I_z$. Then

$$I = \frac{1}{3}(I_x + I_y + I_z) = \frac{1}{3}\iiint \rho \cdot (y^2 + z^2 + x^2 + z^2 + x^2 + y^2) \mathrm{d}V = \frac{2}{3}\rho \int r^2 \mathrm{d}V,$$

where $r^2 = x^2 + y^2 + z^2$ is the distance from point r to the origin. This integral is easy to evaluate in the spherical coordinates, the volume element will be equal to dV= 4ðr^2dr, where r goes from 0 to R. Thus,

$$I = \frac{2}{3}\rho\int_0^R 4\pi r^4\,dr = \frac{2}{3}\rho\cdot 4\pi\frac{R^5}{5} = \frac{m}{\frac{4}{3}\pi R^3}\cdot\frac{8\pi R^5}{15} = \frac{2}{5}mR^2.$$

In mechanics of machines, when designing rotary parts like gears, pulleys, shafts, couplings etc., which are used to transmit torques, the moment of inertia has to be considered.

The moment of inertia is given about an axis and it depends on the shape, density of a rotating element. When considering mechanisms like gear trains, worm and wheel, where there are more than one rotating element, more than one axis of rotation, an equivalent moment of inertia for the system should be found.

Practically when a geared system is enclosed, equivalent moment of inertia can be measured by measuring the angular acceleration for a known torque or theoretically it can be estimated when the masses and dimensions of the rotating elements and shafts are known. In this practical the equivalent moment of inertia of a worm and wheel system is measured using above mention methods.

Moment of Inertia Tensor

In three dimensions, if the axis of rotation is not given, we need to be able to generalize the scalar moment of inertia to a quantity that allows us to compute a moment of inertia about *arbitrary axes*. This quantity is known as the moment of inertia tensor and can be represented as a symmetric positive semi-definite matrix, I. This representation elegantly generalizes the scalar case: The angular momentum vector is related to the rotation velocity vector ù by

$$L = \mathrm{I}\omega$$

and the kinetic energy is given by

$$\frac{1}{2}\omega^{\cdot}\,\mathrm{I}\omega$$

as compared with

$$\frac{1}{2}I\omega^2$$

in the scalar case.

Like the scalar moment of inertia, the moment of inertia tensor may be calculated with respect to any point in space, but for practical purposes, the centre of mass is almost always used. In general, its components are time dependent.

Derivation of the Tensor Components

The distance r of a particle at x from the axis of rotation passing through the origin in the $\hat{n}$ direction is $|x-(x\cdot\hat{n})\hat{n}|$. By using the formula $I = mr^2$ (and some simple vector algebra) it can be seen that the moment of inertia of this particle (about the axis of rotation passing through the origin in the direction)) is $I = m(|x|^2 (\hat{n}\cdot\hat{n}) - (x\cdot\hat{n})^2)$ This is a quadratic form in $\hat{n}$ and, after a bit more algebra, this leads to a tensor formula for the moment of inertia

$$I = m[n_1 n_2 n_3]\begin{bmatrix} y^2+z^2 & -xy & -xz \\ -yx & x^2+z^2 & -yz \\ -zx & -zy & x^2+y^2 \end{bmatrix}\begin{bmatrix} n_1 \\ n_2 \\ n_3 \end{bmatrix}.$$

This is exactly the formula given below for the moment of inertia in the case of a single particle. For multiple particles we need only recall that the moment of inertia is additive in order to see that this formula is correct.

Reduction to Scalar

For any axis $\hat{n}$, represented as a column vector with elements n_i, the scalar form I can be calculated from the tensor form I as

$$I = \hat{n}^{\bullet}\ I\hat{n} = \sum_{j=1}^{3}\sum_{k=1}^{3} n_j I_{jk} n_k$$

The range of both summations correspond to the three Cartesian coordinates. The following equivalent expression avoids the use of transposed vectors which are not supported in maths libraries because internally vectors and their transpose are stored as the same linear array,

$$I = I^{\bullet}\ \hat{n}\cdot\hat{n}$$

However it should be noted that although this equation is mathematically equivalent to the equation above for any matrix, inertia tensors are symmetrical. This means that it can be further simplified to:

$$I = I\hat{n}\cdot\hat{n}$$

Principal Axes of Inertia

By the spectral theorem, since the moment of inertia tensor is real and symmetric, there exists a Cartesian coordinate system in which it is diagonal, having the form

$$\mathrm{I}=\begin{bmatrix} I_1 & 0 & 0 \\ 0 & I_2 & 0 \\ 0 & 0 & I_3 \end{bmatrix}$$

where the coordinate axes are called the principal axes and the constants I_1, I_2 and I_3 are called the principal moments of inertia.

This result was first shown by J. J. Sylvester (1852), and is a form of Sylvester's law of inertia. The principal axis with the highest moment of inertia is sometimes called the figure axis or axis of figure.

When all principal moments of inertia are distinct, the principal axes through centre of mass are uniquely specified. If two principal moments are the same, the rigid body is called a symmetrical top and there is no unique choice for the two corresponding principal axes. If all three principal moments are the same, the rigid body is called a spherical top (although it need not be spherical) and any axis can be considered a principal axis, meaning that the moment of inertia is the same about any axis.

The principal axes are often aligned with the object's symmetry axes. If a rigid body has an axis of symmetry of order m, i.e., is symmetrical under rotations of 360°/m about a given axis, the symmetry axis is a principal axis. When $m > 2$, the rigid body is a symmetrical top. If a rigid body has at least two symmetry axes that are not parallel or perpendicular to each other, it is a spherical top, e.g., a cube or any other Platonic solid. The motion of vehicles is often described about these axes with the rotations called yaw, pitch, and roll. A practical example of this mathematical phenomenon is the routine automotive task of balancing a tire, which basically means adjusting the distribution of mass of a car wheel such that its principal axis of inertia is aligned with the axle so the wheel does not wobble.

Parallel Axis Theorem

Once the moment of inertia tensor has been calculated for rotations about the centre of mass of the rigid body, there is a useful labour-saving method to compute the tensor for rotations offset from the centre of mass.

If the axis of rotation is displaced by a vector R from the centre of mass, the new moment of inertia tensor equals

$$\mathrm{I}^{\text{displaced}} = \mathrm{I}^{\text{center}} + m\left[(\mathrm{R}\cdot\mathrm{R})\mathrm{E}_3 - \mathrm{R}\otimes\mathrm{R}\right]$$

where m is the total mass of the rigid body, E_3 is the 3 × 3 identity matrix, and $\otimes$ is the outer product.

Rotational Symmetry

Using the above equation to express all moments of inertia in terms of integrals of variables either along or perpendicular to the axis of symmetry usually simplifies the calculation of these moments considerably.

Comparison with Covariance Matrix

The moment of inertia tensor about the centre of mass of a 3 dimensional rigid body is related to the covariance matrix of a trivariate random vector whose probability density function is proportional to the pointwise density of the rigid body by:

$$I = n(1_{3\times 3}\,\mathrm{tr}(\Sigma) - \Sigma)$$

where n is the number of points.

The structure of the moment-of-inertia tensor comes from the fact that it is to be used as a bilinear form on rotation vectors in the form

$$\frac{1}{2}\omega^{\cdot}\, I\omega$$

Each element of mass has a kinetic energy of

$$\frac{1}{2}m\,|\,v\,|^2$$

The velocity of each element of mass is $\omega \times r$ where r is a vector from the centre of rotation to that element of mass. The cross product can be converted to matrix multiplication so that

$$\omega \times r = [r]^{\cdot}_{\times}\,\omega$$

and similarly

$$(\omega \times r)^{\cdot} = ([r]_{\times}\omega)^{\cdot} = \omega^{\cdot}\,[r]^{\cdot}_{\times}$$

Thus,

$$|\,v\,|^2 = (\omega \times r)^{\cdot}\,(\omega \times r) = \omega^{\cdot}\,[r]^{\cdot}_{\times}\,[r]_{\times}\omega$$

plugging in the definition of $[\cdot]_{\times}$ the $[r]^{\cdot}_{\times}\,[r]_{\times}$ term leads directly to the structure of the moment tensor.

Truss

In architecture and structural engineering, a truss is a structure comprising one or more triangular units constructed with straight members whose ends are connected at joints referred to as nodes. External forces and reactions to those forces are considered to act only at the nodes and result in forces in the members which are either tensile or compressive forces.

Figure: *Truss bridge for a single-track railway, converted to pedestrian use and pipeline support*

Moments (torques) are explicitly excluded because, and only because, all the joints in a truss are treated as revolutes. A planar truss is one where all the members and nodes lie within a two dimensional plane, while a space truss has members and nodes extending into three dimensions.

Characteristics of Trusses

A truss consists of straight members connected at joints. Trusses are composed of triangles because of the structural stability of that shape and design. A triangle is the simplest geometric figure that will not change shape when the lengths of the sides are fixed. In comparison, both the angles and the lengths of a four-sided figure must be fixed for it to retain its shape.

Planar Truss

The simplest form of a truss is one single triangle. This type of truss is seen in a framed roof consisting of rafters and a ceiling joist. and in other mechanical structures such as bicycles and aircraft. Because of the stability of this shape and the methods of analysis used to calculate the forces within it, a truss composed entirely of triangles is known as a simple truss. The traditional diamond-shape bicycle frame, which utilises two conjoined triangles, is an example of a simple truss.

A planar truss lies in a single plane. Planar trusses are typically used in parallel to form roofs and bridges. The depth of a truss, or the height between the upper and lower chords, is what makes it an efficient structural form. A solid girder or beam of equal strength would have substantial weight and material cost as compared to a truss. For a given span length, a deeper truss will require less material in the chords and greater material in the verticals and diagonals. An optimum depth of the truss will maximize the efficiency.

Space Frame Truss

A space frame truss is a three-dimensional framework of members pinned at their ends. A tetrahedron shape is the simplest space truss, consisting of six members which meet at four joints. Large planar structures may be composed from tetrahedrons with common edges and they are also employed in the base structures of large freestanding power line pylons

Truss Types

Figure : *A large timber Howe truss in a commercial building*

There are two basic types of truss:

- The *pitched truss*, or *common truss*, is characterised by its triangular shape. It is most often used for roof construction. Some common trusses are named according to their *web configuration*. The chord size and web configuration are determined by *span*, *load* and *spacing*.

- The *parallel chord truss*, or *flat truss*, gets its name from its parallel top and bottom chords. It is often used for floor construction.

A combination of the two is a truncated truss, used in hip roof construction. A metal plate-connected wood truss is a roof or floor truss whose wood members are connected with metal connector plates.

Pratt Truss

The Pratt truss was patented in 1844 by two Boston railway engineers, Caleb Pratt and his son Thomas Willis Pratt. The design uses vertical members for compression and horizontal members to respond to tension. What is remarkable about this style is that it remained popular even as wood gave way to iron, and even still as iron gave way to steel. The continued popularity of the Pratt truss is probably due to the fact that the configuration of the members means that longer diagonal members are only in tension for gravity load effects. This allows these members to be used more efficiently, as slenderness effects related to buckling under compression loads (which are compounded by the length of the member) will typically not control the design. Therefore, for given planar truss with a fixed depth, the Pratt configuration is usually the most efficient under static, vertical loading. The Southern Pacific Railroad bridge in Tempe, Arizona is a 393 metre (1,291 foot) long truss bridge built in 1912. The structure is composed of nine Pratt truss spans of varying lengths. The bridge is still in use today. The Wright Flyer used a Pratt truss in its wing construction, as the minimisation of compression member lengths allowed for lower aerodynamic drag.

Bowstring Truss

Named for their shape, bowstring trusses were first used for arched truss bridges, known as tied-arch bridges. Thousands of bowstring trusses were used during World War II for holding up the curved roofs of aircraft hangars and other military buildings. Many variations exist in the arrangements of the members connecting the nodes of the upper arc with those of the lower, straight sequence of members, from nearly isosceles triangles to a variant of the Platt truss.

King Post Truss

Lenticular Truss: Lenticular trusses, patented in 1878 by William Douglas, have the top and bottom chords of the truss arched, forming a lens shape. A lenticular pony truss bridge is a bridge design that involves a lenticular truss extending above and below the roadbed.

Town's Lattice Truss

American architect Ithiel Town designed Town's Lattice Truss as an alternative to heavy-timber bridges. His design, patented in 1820 and 1835, uses easy-to-handle planks arranged diagonally with short spaces in between them.

Vierendeel Truss

***Figure** : A Vierendeel bridge; note the lack of diagonal elements in the primary structure and the way bending loads are carried between elements*

The Vierendeel truss is a truss where the members are not triangulated but form rectangular openings, and is a frame with fixed joints that are capable of transferring and resisting bending moments. Regular trusses comprise members that are commonly assumed to have pinned joints, with the implication that no moments exist at the jointed ends. This style of truss was named after the Belgian engineer Arthur Vierendeel, who developed the design in 1896. Its use for bridges is rare due to higher costs compared to a triangulated truss.

The utility of this type of truss in buildings is that a large amount of the exterior envelope remains unobstructed and can be used for fenestration and door openings. This is preferable to a braced-frame system, which would leave some areas obstructed by the diagonal braces.

Statics of Trusses

A truss that is assumed to comprise members that are connected by means of pin joints, and which is supported at both ends by means of hinged joints or rollers, is described as being statically determinate.

Newton's Laws apply to the structure as a whole, as well as to each node or joint. In order for any node that may be subject to an external load or force to remain static in space, the following conditions must hold: the sums of all (horizontal and vertical) forces, as well as all moments acting about the node equal zero. Analysis of these conditions at each node yields the magnitudee may be compression or tension forces.

Trusses that are supported at more than two positions are said to be statically indeterminate, and the application of Newton's Laws alone is not sufficient to determine the member forces. In order for a truss with pin-connected members to be stable, it must be entirely composed of triangles. In mathematical terms, we have the following necessary condition for stability:

$$m \geq 2j - r \qquad \text{(a)}$$

where m is the total number of truss members, j is the total number of joints and r is the number of reactions (equal to 3 generally) in a 2 dimensional structure.

When $m = 2j - 3$, the truss is said to be *statically determinate*, because the $(m+3)$ internal member forces and support reactions can then be completely determined by $2j$ equilibrium equations, once we know the external loads and the geometry of the truss. Given a certain number of joints, this is the minimum number of members, in the sense that if any member is taken out (or fails), then the truss as a whole fails. While the relation (a) is necessary, it is not sufficient for stability, which also depends on the truss geometry, support conditions and the load carrying capacity of the members.

Some structures are built with more than this minimum number of truss members. Those structures may survive even when some of the members fail. Their member forces depend on the relative stiffness of the members, in addition to the equilibrium condition described.

Analysis of Trusses

Because the forces in each of its two main girders are essentially planar, a truss is usually modelled as a two-dimensional plane frame. If there are significant out-of-plane forces, the structure must be modelled as a three-dimensional space. The analysis of trusses often assumes that loads are applied to joints only and not at intermediate points along the members. The weight of the members is often insignificant compared to the applied loads and so is often omitted. If required, half of the weight of each member may be applied to its two end joints. Provided the members are long and slender, the moments transmitted through the joints are negligible and they can be treated as "hinges" or 'pin-joints'. Every member of the truss is then in pure compression or pure tension – shear, bending moment, and other more complex stresses are all practically zero. This makes trusses easier to analyse. This also makes trusses physically stronger than other ways of arranging material – because nearly every material can hold a much larger load in tension and compression than in shear, bending, torsion, or other kinds of force.

Structural analysis of trusses of any type can readily be carried out using a matrix method such as the direct stiffness method, the flexibility method or the finite element method.

Forces in Members

On the right is a simple, statically determinate flat truss with 9 joints and (2 x 9) " 3 = 15 members. External loads are concentrated in the outer joints. Since this is a symmetrical truss with symmetrical vertical loads, it is clear to see that the reactions at A and B are equal, vertical and half the total load.

The internal forces in the members of the truss can be calculated in a variety of ways including the graphical methods:

- *Cremona diagram*
- *Culmann diagram*
- *the analytical* Ritter *method (method of sections).*

Design of Members

A truss can be thought of as a beam where the web consists of a series of separate members instead of a continuous plate. In the truss, the lower horizontal member (the bottom chord) and the upper horizontal member (the top chord) carry tension and compression, fulfilling the same function as the flanges of an I-beam. Which chord carries tension and which carries compression depends on the overall direction of bending. In the truss pictured above right, the bottom chord is in tension, and the top chord in compression.

The diagonal and vertical members form the truss web, and carry the shear force. Individually, they are also in tension and compression, the exact arrangement of forces is depending on the type of truss and again on the direction of bending. In the truss shown above right, the vertical members are in tension, and the diagonals are in compression.

In addition to carrying the static forces, the members serve additional functions of stabilising each other, preventing buckling. In the picture to the right, the top chord is prevented from buckling by the presence of bracing and by the stiffness of the web members.

The inclusion of the elements shown is largely an engineering decision based upon economics, being a balance between the costs of raw materials, off-site fabrication, component transportation, on-site erection, the availability of machinery and the cost of labour. In other cases the appearance of the structure may take on greater importance and so influence the design decisions beyond mere matters of economics. Modern materials such as prestressed concrete and fabrication methods, such as automated welding, have significantly influenced the design of modern bridges.

Once the force on each member is known, the next step is to determine the cross section of the individual truss members. For members under tension the cross-sectional area A can be found using $A = F \times$ ã / ó$_y$, where F is the force in the member, ã is a safety factor (typically 1.5 but depending on building codes) and ó$_y$ is the yield tensile strength of the steel used.

The members under compression also have to be designed to be safe against buckling. The weight of a truss member depends directly on its cross section—that weight partially determines how strong the other members of the truss need to be. Giving one member a larger cross section than on a previous iteration requires giving other members a larger cross section as well, to hold the greater weight of the first member—one needs to go through another iteration to find exactly how much greater the other members need to be. Sometimes the designer goes through several iterations of the design process to converge on the "right" cross section for each member. On the other hand, reducing the size of one member from the previous iteration merely makes the other members have a larger (and more expensive) safety factor than is technically necessary, but doesn't *require* another iteration to find a buildable truss.

The effect of the weight of the individual truss members in a large truss, such as a bridge, is usually insignificant compared to the force of the external loads.

Design of Joints

After determining the minimum cross section of the members, the last step in the design of a truss would be detailing of the bolted joints, e.g., involving shear of the bolt connections used in the joints, see also shear stress. Based of the needs of the project, truss internal connections (joints) can be designed as rigid, semi rigid, or hinged. Rigid connections can allow transfer of bending moments leading to development of secondary bending moments in the members.

Applications

Post Frame Structures: Component connections are critical to the structural integrity of a framing system. In buildings with large, clearspan wood trusses, the most critical connections are those between the truss and its supports. In addition to gravity-induced forces (a.k.a. bearing loads), these connections must resist shear forces acting perpendicular to the plane of the truss and uplift forces due to wind. Depending upon overall building design, the connections may also be required to transfer bending moment.

Wood posts enable the fabrication of strong, direct, yet inexpensive connections between large trusses and walls. Exact details for post-to-truss connections vary from designer to designer, and may be influenced by post type. Solid-sawn timber and glulam posts are generally notched to form a truss bearing surface. The truss is rested on the notches and bolted into place. A special plate/bracket may be added to increase connection load transfer capabilities. With mechanically-laminated posts, the truss may rest on a shortened outer-ply or on a shortened inner-ply. The later scenario places the bolts in double shear and is a very effective connection.

Structural Analysis

Structural analysis is the determination of the effects of loads on physical structures and their components. Structures subject to this type of analysis include all that must withstand loads, such as buildings, bridges, vehicles, machinery, furniture, attire, soil strata, prostheses and biological tissue. Structural analysis incorporates the fields of applied mechanics, materials science and applied mathematics to compute a structure's deformations, internal forces, stresses, support reactions, accelerations, and stability. The results of the analysis are used to verify a structure's fitness for use, often saving physical tests. Structural analysis is thus a key part of the engineering design of structures.

Structures and Loads

A structure refers to a body or system of connected parts used to support a load. Important examples related to Civil Engineering include buildings, bridges, and towers; and in other branches of engineering, ship and aircraft frames, tanks, pressure vessels, mechanical systems, and electrical supporting structures are important. In order to design a structure, one must serve a specified function for public use, the engineer must account for its safety, aesthetics, and serviceability, while taking into consideration economic and environmental constraints. Other branches of engineering work on a wide variety of nonbuilding structures.

Classification of Structures

It is important for a structural engineer to recognize the various types of elements composing a structure and to be able to classify structures as to their form and function. Some of the structural elements are tie rods, rod, bar, angle, channel, beams, and columns. Combination of structural elements and the materials from which they are composed is referred to as a structural system. Each system is constructed of one or more basic types of structures such as Trusses, Cables and Arches, Frames, and Surface Structures.

Loads

Once the dimensional requirement for a structure have been defined, it becomes necessary to determine the loads the structure must support. In order to design a structure, it is therefore necessary to first specify the loads that act on it. The design loading for a structure is often specified in building codes. There are two types of codes: general building codes and design codes, engineer must satisfy all the codes requirements for a reliable structure. There are two types of loads that structure engineering must encounter in the design. First type of load is called Dead loads that consist of the weights of the various structural members and the weights of any objects that are permanently attached to the structure. For example, columns, beams, girders, the floor slab, roofing, walls, windows, plumbing, electrical fixtures, and other miscellaneous attachments. Second type of load is Live Loads which vary in their magnitude and location. There are many different types of live loads like building loads, highway bridge Loads, railroad bridge Loads, impact loads, wind loads, snow loads, earthquake loads, and other natural loads.

Analytical Methods

To perform an accurate analysis a structural engineer must determine such information as structural loads, geometry, support conditions, and materials properties. The results of such an analysis typically include support reactions, stresses and displacements. This information is then compared to criteria that indicate the conditions of failure. Advanced structural analysis may examine dynamic response, stability and nonlinear behaviour.

There are three approaches to the analysis: the mechanics of materials approach (also known as strength of materials), the elasticity theory approach (which is actually a special case of the more general field of continuum mechanics), and the finite element approach. The first two make use of analytical formulations which apply mostly to simple linear elastic modells, lead to closed-form solutions, and can often be solved by hand. The finite element approach is actually a numerical method for solving differential equations generated by theories of mechanics such as elasticity theory and strength of materials. However, the finite-element method depends heavily on the processing power of computers and is more applicable to structures of arbitrary size and complexity. Regardless of approach, the formulation is based on the same three fundamental relations: equilibrium, constitutive, and compatibility. The solutions are approximate when any of these relations are only approximately satisfied, or only an approximation of reality.

Limitations

Each method has noteworthy limitations. The method of mechanics of materials is limited to very simple structural elements under relatively simple loading conditions. The structural elements and loading conditions allowed, however, are sufficient to solve many useful engineering problems. The theory of elasticity allows the solution of structural elements of general geometry under general loading conditions, in principle. Analytical solution, however, is limited to relatively simple cases. The solution of elasticity problems also requires the solution of a system of partial differential equations, which is considerably more mathematically demanding than the solution of mechanics of materials problems, which require at most the solution of an ordinary differential equation. The finite element method is perhaps the most restrictive and most useful at the same time. This method itself relies upon other structural theories (such as the other two discussed here) for equations to solve. It does, however, make it generally possible to solve these equations, even with highly complex geometry and loading conditions, with the restriction that there is always some numerical error. Effective and reliable use of this method requires a solid understanding of its limitations.

Strength of Materials Methods (Classical Methods)

The simplest of the three methods here discussed, the mechanics of materials method is available for simple structural members subject to specific loadings such as axially loaded bars, prismatic beams in a state of pure bending, and circular shafts subject to torsion. The solutions can under certain conditions be superimposed using the superposition principle to analyse a member undergoing combined loading. Solutions for special cases exist for common structures such as thin-walled pressure vessels. For the analysis of entire systems, this approach can be used in conjunction with statics, giving rise to the *method of sections* and *method of joints* for truss analysis, moment distribution method for small rigid frames, and *portal frame* and *cantilever method* for large rigid frames. Except for moment distribution, which came into use in the 1930's, these methods were developed in their current forms in the second half of the nineteenth century. They are still used for small structures and for preliminary design of large structures.

The solutions are based on linear isotropic infinitesimal elasticity and Euler-Bernoulli beam theory. In other words, they contain the assumptions (among others) that the materials in question are elastic, that stress is related linearly to strain, that the material (but not the structure) behaves identically regardless of direction of the applied load, that all deformations are small, and that beams are long relative to their depth. As with any simplifying assumption in engineering, the more the model strays from reality, the less useful (and more dangerous) the result.

Elasticity Methods

Elasticity methods are available generally for an elastic solid of any shape. Individual members such as beams, columns, shafts, plates and shells may be modeled. The solutions are derived from the equations of linear elasticity. The equations of elasticity are a system of 15 partial differential equations. Due to the nature of the mathematics involved, analytical solutions may only be produced for relatively simple geometries. For complex geometries, a numerical solution method such as the finite element method is necessary. Many of the developments in the mechanics of materials and elasticity approaches have been expounded or initiated by Stephen Timoshenko.

Methods Using Numerical Approximation

It is common practice to use approximate solutions of differential equations as the basis for structural analysis. This is usually done using numerical approximation techniques. The most commonly used numerical approximation in structural analysis is the Finite Element Method.

The finite element method approximates a structure as an assembly of elements or components with various forms of connection between them. Thus, a continuous system such as a plate or shell is modeled as a discrete system with a finite number of elements interconnected at finite number of nodes. The behaviour of individual elements is characterised by the element's stiffness or flexibility relation, which altogether leads to the system's stiffness or flexibility relation. To establish the element's stiffness or flexibility relation, we can use the *mechanics of materials* approach for simple one-dimensional bar elements, and the *elasticity approach* for more complex two- and three-dimensional elements. The analytical and computational development are best effected throughout by means of matrix algebra.

Early applications of matrix methods were for articulated frameworks with truss, beam and column elements; later and more advanced matrix methods, referred to as "finite element analysis," model an entire structure with one-, two-, and three-dimensional elements and can be used for articulated systems together with continuous systems such as a pressure vessel, plates, shells, and three-dimensional solids. Commercial computer software for structural analysis typically uses matrix finite-element analysis, which can be further classified into two main approaches: the displacement or stiffness method and the force or flexibility method. The stiffness method is the most popular by far thanks to its ease of implementation as well as of formulation for advanced applications. The finite-element technology is now sophisticated enough to handle just about any system as long as sufficient computing power is available. Its applicability includes, but is not limited to, linear and nonlinear analysis, solid and fluid interactions, materials that are isotropic, orthotropic, or anisotropic, and external effects that are static, dynamic, and environmental factors. This, however, does not imply that the computed solution will automatically be reliable because much depends on the model and the reliability of the data input.

Applied Element Method

The Applied Element Method (AEM) is a numerical analysis used in predicting the continuum and discrete behaviour of structures. The modelling method in AEM adopts the concept of discrete cracking allowing it to automatically track structural collapse behaviour passing through all stages of loading: elastic, crack initiation and propagation in tension-weak materials, reinforcement yield, element separation, element contact and collision, as well as collision with the ground and adjacent structures.

History

Exploration of the approach employed in the applied element method began in 1995 at the University of Tokyo as part of Dr.Hatem Tagel-Din's research studies. The term itself "Applied Element Method," however was first coined in 2000 in a paper called "Applied element method for structural analysis: Theory and application for linear Materials.". Since then AEM has been the subject of research by a number of academic institutions and the driving factor in real-world applications. Research has verified its accuracy for: elastic analysis; crack initiation and propagation; estimation of failure loads at reinforced concrete structures; reinforced concrete structures under cyclic loading; buckling and post-buckling behaviour; nonlinear dynamic analysis of structures subjected to severe earthquakes; fault-rupture propagation; nonlinear behaviour of brick structures; and the analysis of glass reinforced polymers (GFRP) walls under blast loads.

Technical Discussion

In AEM, the structure is divided virtually and modeled as an assemblage of relatively small elements. The elements are then connected through a set of normal and shear springs located at contact points distributed along the element faces. Normal and shear springs are responsible for the transfer of normal and shear stresses from one element to the next.

Element Generation and Formulation

The modelling of objects in AEM is very similar to modelling objects in FEM. Each object is divided into a series of elements connected together and forming a mesh. The main difference between AEM and FEM, however, is how the elements are joined together. In AEM the elements are connected by a series of nonlinear springs representing the material behaviour.

There are three types of springs used in AEM:

- Matrix Springs: Matrix springs connect two elements together representing the main material properties of the object.
- Reinforcing Bar Springs: Reinforcement springs are used to implicitly represent additional reinforcement bars running through the object without adding additional elements to the analysis.
- Contact Springs: Contact Springs are generated when two elements collide with each other or the ground. When this occurs three springs are generated (Shear Y, Shear X and Normal).

Automatic Element Separation

When the average strain value at the element face reaches the separation strain, all springs at this face are removed and elements are no longer connected until a collision occurs, at which point they collide together as rigid bodies.

Separation strain represents the strain at which adjacent elements are totally separated at the connecting face. This parametre is not available in the elastic material model. For concrete, all springs between the adjacent faces including reinforcement bar springs are cut. If the elements meet again, they will behave as two different rigid bodies that have now contacted each other. For steel, the bars are cut if the stress point reaches ultimate stress or if the concrete reaches the separation strain.

Automatic Element Contact/Collision

Contact or collision is detected without any user intervention. Elements are able to separate, contract and/or make contact with other elements. In AEM three contact methods include Corner-to-Face, Edge-to-Edge, and Corner-to-Ground.

Stiffness Matrix

The spring stiffness in a 2D model can be calculated from the following equations:

$$K_n = \frac{E \cdot T \cdot d}{a}$$

$$K_s = \frac{G \cdot T \cdot d}{a}$$

Where d is the distance between springs, T is the thickness of the element, a is the length of the representative area, E is the Young's modulus, and G is the shear modulus of the material. The above equation's indicate that each spring represents the stiffness of an area ($T \cdot d$) within the length of the studied material.

To model reinforcement bars embedded in concrete, a spring is placed inside the element at the location of the bar; the area ($T \cdot d$) is replaced by the actual cross section area of the reinforcement bar. Similar to modelling embedded steel sections, the area ($T \cdot d$) may be replaced by the area of the steel section represented by the spring.

Although the element motion moves as a rigid body, its internal deformations are represented by the spring deformation around each element. This means the element shape does not change during analysis, but the behaviour of assembly of elements is deformable. The two elements are assumed to be connected by only one pair of normal and shear springs.

To have a general stiffness matrix, the locations of element and contact springs are assumed in a general position. The stiffness matrix components corresponding to each degree of freedom are determined by assuming a unit displacement in the studied direction and by determining forces at the centroid of each element. The 2D element stiffness matrix size is 6 x 6; the components of the upper left quarter of the stiffness matrix are shown below:

$$\begin{bmatrix} \sin^2(\theta+\alpha)K_n & -K_n\sin(\theta+\alpha)\cos(\theta+\alpha) & \cos(\theta+\alpha)K_sL\sin(\alpha) \\ +\cos^2(\theta+\alpha)K_s & +K_s\sin(\theta+\alpha)\cos(\theta+\alpha) & -\sin(\theta+\alpha)K_nL\cos(\alpha) \\ -K_n\sin(\theta+\alpha)\cos(\theta+\alpha) & \sin^2(\theta+\alpha)K_s & \cos(\theta+\alpha)K_nL\cos(\alpha) \\ -K_s\sin(\theta+\alpha)\cos(\theta+\alpha) & +\cos^2(\theta+\alpha)K_n & +\sin(\theta+\alpha)K_sL\sin(\alpha) \\ \cos(\theta+\alpha)K_sL\sin(\alpha) & \cos(\theta+\alpha)K_nL\cos(\alpha) & L^2\cos^2(\alpha)K_n \\ -\sin(\theta+\alpha)K_nL\cos(\alpha) & +\sin(\theta+\alpha)K_sL\sin(\alpha) & +L^2\sin^2(\alpha)K_s \end{bmatrix}$$

The stiffness matrix depends on the contact spring stiffness and the spring location. The stiffness matrix is for only one pair of contact springs. However, the global stiffness matrix is determined by summing up the stiffness matrices of individual pairs of springs around each element. Consequently, the developed stiffness matrix has total effects from all pairs of springs, according to the stress situation around the element. This technique can be used in both load and displacement control cases. The 3D stiffness matrix may be deduced similarly.

Applications

The applied element method is currently being used in the following applications:

- Structural vulnerability assessment
  - o Progressive collapse
  - o Blast analysis
  - o Impact analysis
  - o Seismic analysis
- Forensic engineering
- Performance based design
- Demolition analysis
- Glass performance analysis
- Visual effects.

Moment Distribution Method

The moment distribution method (not to be confused with moment redistribution) is a structural analysis method for statically indeterminate beams and frames developed by Hardy Cross. It was published in 1930 in an ASCE journal. The method only accounts for flexural effects and ignores axial and shear effects. From the 1930's until computers began to be widely used in the design and analysis of structures, the moment distribution method was the most widely practised method.

In the moment distribution method, every joint of the structure to be analysed is fixed so as to develop the *fixed-end moments*. Then each fixed joint is sequentially released and the fixed-end moments (which by the time of release are not in equilibrium) are distributed to adjacent members until equilibrium is achieved. The moment distribution method in mathematical terms can be demonstrated as the process of solving a set of simultaneous equations by means of iteration. The moment distribution method falls into the category of displacement method of structural analysis.

Fixed End Moments

$$M^f_{AB} = \frac{Pb^2a}{L^2} = \frac{10\times 7^2\times 3}{10^2} = +14.700\ kN{\cdot}m$$

$$M^f_{BA} = -\frac{Pa^2b}{L^2} = -\frac{10\times 3^2\times 7}{10^2} = -6.300\ kN{\cdot}m$$

$$M^f_{BC} = \frac{qL^2}{12} = \frac{1\times 10^2}{12} = +8.333\ kN{\cdot}m$$

$$M^f_{CB} = -\frac{qL^2}{12} = -\frac{1\times 10^2}{12} = -8.333\ kN{\cdot}m$$

$$M^f_{CD} = \frac{PL}{8} = \frac{10\times 10}{8} = +12.500\ kN{\cdot}m$$

$$M^f_{DC} = -\frac{PL}{8} = -\frac{10\times 10}{8} = -12.500\ kN{\cdot}m$$

Flexural Stiffness and Distribution Factors

The flexural stiffness of members AB, BC and CD are $\frac{3EI}{L}$ $\frac{4\times 2EI}{L}$ and $\frac{4EI}{L}$ respectively. Therefore:

$$D_{BA} = \frac{\frac{3EI}{L}}{\frac{3EI}{L} + \frac{4 \times 2EI}{L}} = \frac{\frac{3}{10}}{\frac{3}{10} + \frac{8}{10}} = 0.2727$$

$$D_{BC} = \frac{\frac{4 \times 2EI}{L}}{\frac{3EI}{L} + \frac{4 \times 2EI}{L}} = \frac{\frac{8}{10}}{\frac{3}{10} + \frac{8}{10}} = 0.7273$$

$$D_{CB} = \frac{\frac{4 \times 2EI}{L}}{\frac{4 \times 2EI}{L} + \frac{4EI}{L}} = \frac{\frac{8}{10}}{\frac{8}{10} + \frac{4}{10}} = 0.6667$$

$$D_{CD} = \frac{\frac{4EI}{L}}{\frac{4 \times 2EI}{L} + \frac{4EI}{L}} = \frac{\frac{4}{10}}{\frac{8}{10} + \frac{4}{10}} = 0.3333$$

The distribution factors of joints A and D are D_{AB}=1 and D_{DC}=0.

Result

- Moments at joints determined by the moment distribution method

$$M_A = 0 \; kN{\cdot}m$$

$$M_B = -11.569 \; kN{\cdot}m$$

$$M_C = -10.186 \; kN{\cdot}m$$

$$M_D = -13.657 \; kN{\cdot}m$$

The conventional engineer's sign convention is used here, i.e. positive moments cause elongation at the bottom part of a beam member. For comparison purposes, the following are the results generated using a matrix method. Note that in the analysis above, the iterative process was carried to >0.01 precision. The fact that the matrix analysis results and the moment distribution analysis results match to 0.001 precision is mere coincidence.

- Moments at joints determined by the matrix method

$$M_A = 0 \; kN{\cdot}m$$

$$M_B = -11.569 \; kN{\cdot}m$$

$$M_C = -10.186 \; kN{\cdot}m$$

$$M_D = -13.657 \; kN{\cdot}m$$

The complete shear and bending moment diagrams are as shown. Note that the moment distribution method only determines the moments at the joints. Developing complete bending moment diagrams require additional calculations using the determined joint moments and internal section equilibrium.

- SFD and BMD

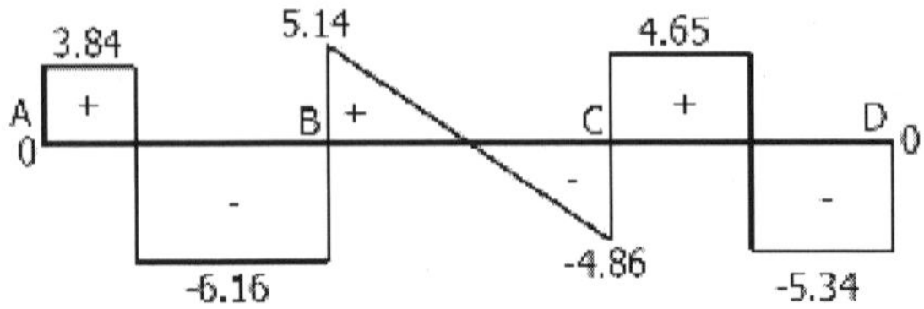

Shear force diagram

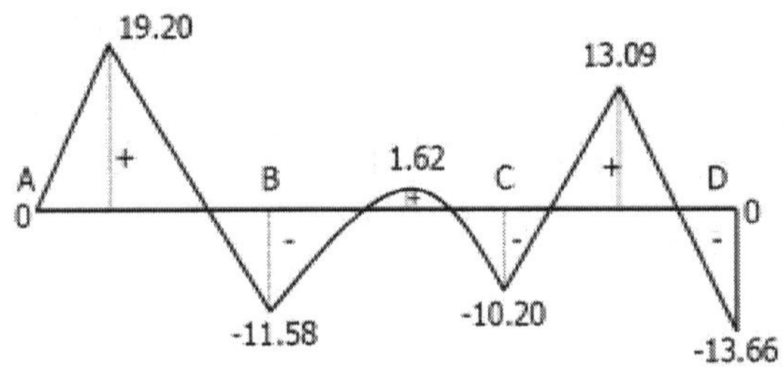

Bending moment diagram

Ground-structure Interaction

Design

The interaction between soil (ground) and structure consists of an exchange of mutual stress between the structure itself and the foundations ground. This means that the movement of the whole ground-structure system is influenced by the type of ground as well as by the type of structure. The main types of foundations, based upon several building characteristics, are:

- Isolated plinths (currently not feasible)
- Plinths connected by foundations beams
- Reverse beams
- A plate (used for low-quality grounds).

The filing of foundations grounds takes place according to the mechanical properties of the grounds themselves: in Italy, for instance, according to the new earthquake-proof norm – Ordinance 3274/2003 – you can identify the following categories:

- Category A: homogeneous rock formations
- Category B: compact granular or clayey soil
- Category C: quite compact granular or clayey soil
- Category D: not much compact granular or clayey soil
- Category E: alluvial surface layer grounds (very low quality soil).

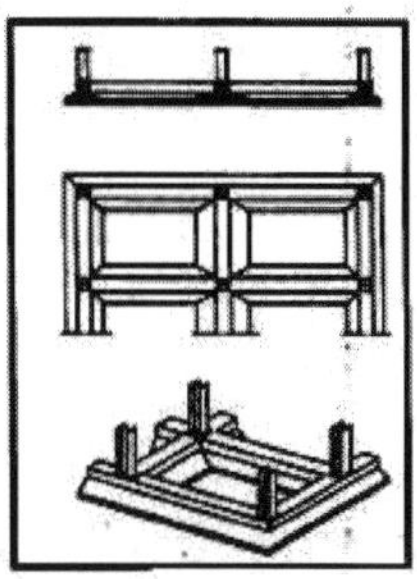

Reverse beams

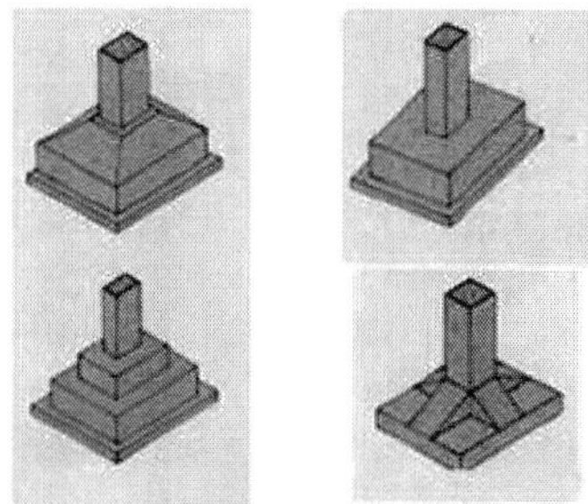

Isolated plinths

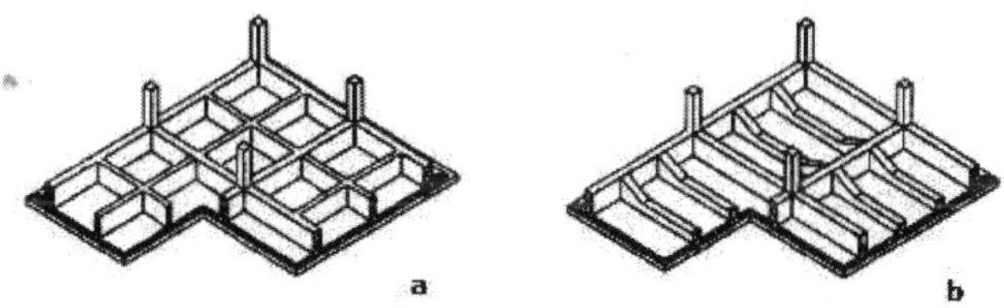

a b

Plate

The type of foundations is selected according to the type of ground; for instance, in the case of homogeneous rock formations connected plinths are selected, while in the case of very low quality grounds plates are chosen.

Both grounds and structures can be more or less deformable; their combination can or cannot cause the amplification of the seismic effects on the structure. Ground, in fact, is a filter with respect to all the main seismic waves, as stiffer soil fosters high-frequency seismic waves while less compact soil accommodates lower frequency waves. Therefore, a stiff building, characterised by a high fundamental frequency, suffers amplified damage when built on stiff ground and then subjected to higher frequencies. For instance, suppose there are two buildings that share the same high stiffness. They stand on two different soil types: the first, stiff and rocky—the second, sandy and deformable. If subjected to the same seismic event, the building on the stiff ground suffers greater damage.

The second interaction effect, tied to mechanical properties of soil, is about the lowering (sinking) of foundations, worsened by the seismic event itself, especially about less compact grounds. This phenomenon is called soil liquefaction.

Mitigation

The methods most used to mitigate the problem of the ground-structure interaction consist of the employment of the before-seen isolation systems and of some ground brace techniques, which are adopted above all on the low-quality ones (categories D and E). The most diffused techniques are the jet-grouting technique and the pilework technique.

The jet-grouting technique consists of injecting in the subsoil some liquid concrete by means of a drill. When this concrete hardens it forms a sort of column that consolidates the surrounding soil. This process is repeated on all areas of the structure. The pilework technique consists of using piles, which, once inserted in the ground, support the foundation and the building above, by moving the loads or the weights towards soil layers that are deeper and therefore more compact and movement-resistant.

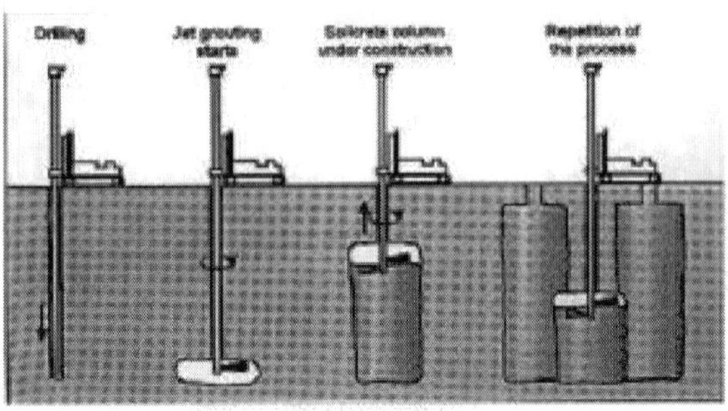

Jet-grouting technique

Pilework technique

Centroid

In geometry, the centroid, geometric centre, or barycentre of a plane figure or two-dimensional shape X is the intersection of all straight lines that divide X into two parts of equal moment about the line. Informally, it is the "average" (arithmetic mean) of all points of X. The definition extends to any object X in n-dimensional space: its centroid is the intersection of all hyperplanes that divide X into two parts of equal moment.

In physics, the word centroid means the geometric centre of the object's shape, as above, but barycentre may also mean its physical centre of mass or the centre of gravity, depending on the context. Informally, the centre of mass (and centre of gravity in a uniform gravitational field) is the average of all points, weighted by the local density or specific weight. If a physical object has uniform density, then its centre of mass is the same as the centroid of its shape. In geography, the centroid of a region of the Earth's surface, projected radially onto said surface, is known as its geographical centre.

Properties

The geometric centroid of a convex object always lies in the object. A non-convex object might have a centroid that is outside the figure itself. The centroid of a ring or a bowl, for example, lies in the object's central void. If the centroid is defined, it is a fixed point of all isometries in its symmetry group. In particular, the geometric centroid of an object lies in the intersection of all its hyperplanes of symmetry. The centroid of many figures (regular polygon, regular polyhedron, cylinder, rectangle, rhombus, circle, sphere, ellipse, ellipsoid, superellipse, superellipsoid, etc.) can be determined by this principle alone. In particular, the centroid of a parallelogram is the meeting point of its two diagonals. This is not true for other quadrilaterals. For the same reason, the centroid of an object with translational symmetry is undefined (or lies outside the enclosing space), because a translation has no fixed point.

Locating the Centroid

Plumb Line Method: The centroid of a uniform two-dimensional lamina, such as (a) below, may be determined, experimentally, by using a plumbline and a pin to find the centre of mass of a thin body of uniform density having the same shape. The body is held by the pin inserted at a point near the body's perimetre, in such a way that it can freely rotate around the pin; and the plumb line is dropped from the pin (b). The position of the plumbline is traced on the body. The experiment is repeated with the pin inserted at a different point of the object. The intersection of the two lines is the centroid of the figure (c).

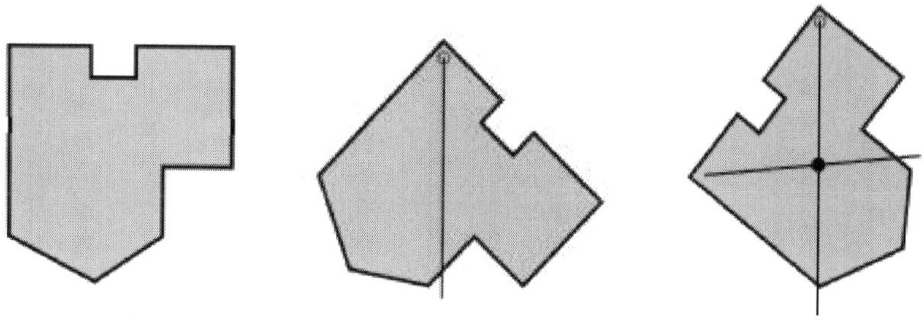

(a) *(b)* *(c)*

This method can be extended (in theory) to concave shapes where the centroid lies outside the shape, and to solids (of uniform density), but the positions of the plumb lines need to be recorded by means other than drawing.

Balancing Method

For convex two-dimensional shapes, the centroid can be found by balancing the shape on a smaller shape, such as the top of a narrow cylinder. The centroid occurs somewhere within the range of contact between the two shapes. In principle, progressively narrower cylinders can be used to find the centroid to arbitrary accuracy. In practice air currents make this unfeasible. However, by marking the overlap range from multiple balances, one can achieve a considerable level of accuracy.

Of a finite set of points

The centroid of a finite set of k points $\mathrm{x}_1, \mathrm{x}_2, \ldots, \mathrm{x}_k$ in $\square^n$ is

$$C = \frac{\mathrm{x}_1 + \mathrm{x}_2 + \cdots + \mathrm{x}_k}{k}$$

By Geometric Decomposition

The centroid of a plane figure X can be computed by dividing it into a finite number of simpler figures $X_1, X_2, \ldots, X_n$, computing the centroid C_i and area A_i of each part, and then computing

$$C = \frac{\sum C_i A_i}{\sum A_i}.$$

Holes in the figure X, overlaps between the parts, or parts that extend outside the figure can all be handled using negative areas A_i. Namely, the measures A_i should be taken with positive and negative signs in such a way that the sum of the signs of A_i for all parts that enclose a given point p is 1 if p belongs to X, and 0 otherwise.

For example, the figure below (a) is easily divided into a square and a triangle, both with positive area; and a circular hole, with negative area (b).

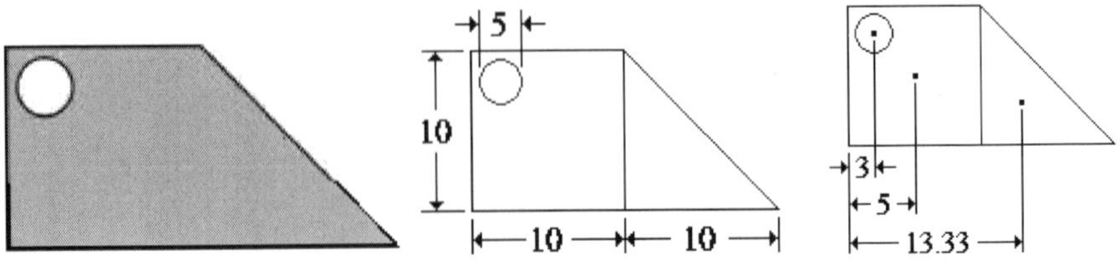

(a) (b) (c)

The centroid of each part can be found in any list of centroids of simple shapes (c). Then the centroid of the figure is the weighted average of the three points. The horizontal position of the centroid, from the left edge of the figure is

$$x = \frac{5\times 10^2 + 13.33\times \frac{1}{2}10^2 - 3\times \pi 2.5^2}{10^2 + \frac{1}{2}10^2 - \pi 2.5^2} \approx 8.5 \text{ units}$$

The same formula holds for any three-dimensional objects, except that each A_i should be the volume of X_i, rather than its area. It also holds for any subset of $\mathbb{R}^d$, for any dimension d, with the areas replaced by the d-dimensional measures of the parts.

By Integral Formula

The centroid of a subset X of $\mathbb{R}^n$ can also be computed by the integral

$$C = \frac{\int x g(x)\, dx}{\int g(x)\, dx}$$

where the integral is taken over the whole space $\mathbb{R}^n$, and g is the characteristic function of the subset, which is 1 inside X and 0 outside it. Note that the denominator is simply the measure of the set X (However, this formula cannot be applied if the set X has zero measure, or if either integral diverges.)

Another formula for the centroid is

$$C_k = \frac{\int z S_k(z)\, dz}{\int S_k(z)\, dz}$$

where C_k is the kth coordinate of C, and $S_k(z)$ is the measure of the intersection of X with the hyperplane defined by the equation $x_k = z$. Again, the denominator is simply the measure of X.

For a plane figure, in particular, the barycentre coordinates are:

$$C_x = \frac{\int x S_y(x)\, dx}{A}$$

$$C_y = \frac{\int y S_x(y)\, dy}{A}$$

where A is the area of the figure X; $S_y(x)$ is the length of the intersection of X with the vertical line at abscissa x; and $S_x(y)$ is the analoguous quantity for the swapped axes.

Of an L-shaped Object

This is a method of determining the centroid of an L-shaped object.

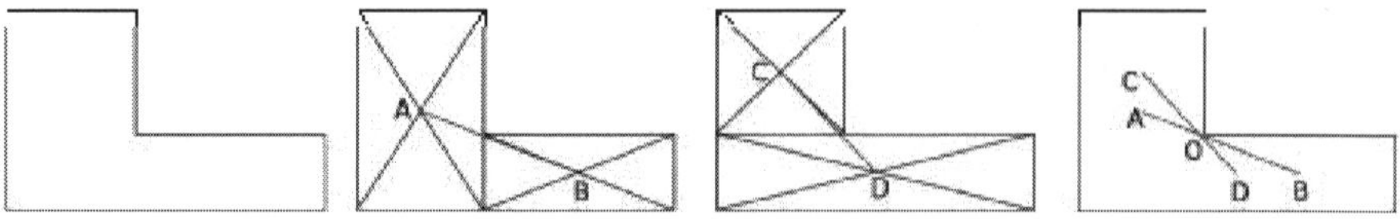

1. Divide the shape into two rectangles, as shown in fig 2. Find the centroids of these two rectangles by drawing the diagonals. Draw a line joining the centroids. The centroid of the shape must lie on this line AB.
2. Divide the shape into two other rectangles, as shown in fig 3. Find the centroids of these two rectangles by drawing the diagonals. Draw a line joining the centroids. The centroid of the L-shape must lie on this line CD.
3. As the centroid of the shape must lie along AB and also along CD, it is obvious that it is at the intersection of these two lines, at O. The point O might *not* lie inside the L-shaped object.

Of Triangle and Tetrahedron

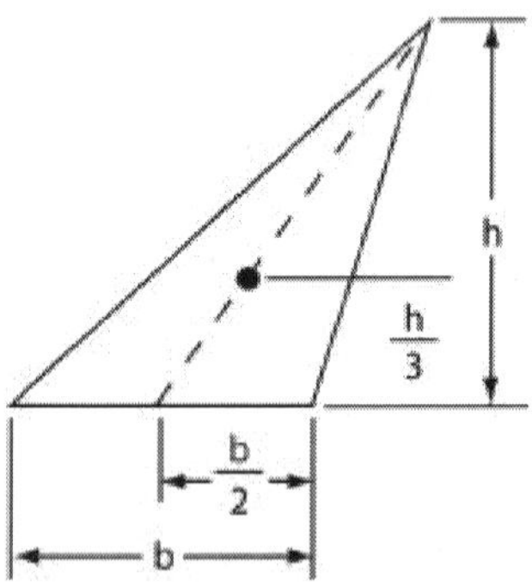

The centroid of a triangle is the point of intersection of its medians (the lines joining each vertex with the midpoint of the opposite side). The centroid divides each of the medians in the ratio 2:1, which is to say it is located S! of the perpendicular distance between each side and the opposing point. Its Cartesian coordinates are the means of the coordinates of the three vertices. That is, if the three vertices are $a = (x_a, y_a)$, $b = (x_b, y_b)$, and $c = (x_c, y_c)$, then the centroid is

$$C=\frac{1}{3}(a+b+c)=\left(\frac{1}{3}(x_a+x_b+x_c),\ \frac{1}{3}(y_a+y_b+y_c)\right).$$

The centroid is therefore at $\left(\frac{1}{3},\frac{1}{3},\frac{1}{3}\right)$ in barycentric coordinates.

The centroid is also the physical centre of mass if the triangle is made from a uniform sheet of material; or if all the mass is concentrated at the three vertices, and evenly divided among them. On the other hand, if the mass is distributed along the triangle's perimetre, with uniform linear density, the centre of mass may not coincide with the geometric centroid. The area of the triangle is 1.5 times the length of any side times the perpendicular distance from the side to the centroid. Similar results hold for a tetrahedron: its centroid is the intersection of all line segments that connect each vertex to the centroid of the opposite face. These line segments are divided by the centroid in the ratio 3:1. The result generalizes to any n-dimensional simplex in the obvious way. If the set of vertices of a simplex is $v_0,\ldots,v_n$, then considering the vertices as vectors, the centroid is

$$C=\frac{1}{n+1}\sum_{i=0}^{n}v_i.$$

The geometric centroid coincides with the centre of mass if the mass is uniformly distributed over the whole simplex, or concentrated at the vertices as n equal masses.

The isogonal conjugate of a triangle's centroid is its symmedian point.

Centroid of Polygon

The centroid of a non-self-intersecting closed polygon defined by n vertices (x_0,y_0), (x_1,y_1),..., $(x_{n"1},y_{n"1})$ is the point (C_x, C_y), where

$$C_x=\frac{1}{6A}\sum_{i=0}^{n-1}(x_i+x_{i+1})(x_i\,y_{i+1}-x_{i+1}\,y_i)$$

$$C_y=\frac{1}{6A}\sum_{i=0}^{n-1}(y_i+y_{i+1})(x_i\,y_{i+1}-x_{i+1}\,y_i)$$

and where A is the polygon's signed area,

$$A=\frac{1}{2}\sum_{i=0}^{n-1}(x_i\,y_{i+1}-x_{i+1}\,y_i)$$

In these formulas, the vertices are assumed to be numbered in order of their occurrence along the polygon's perimetre, and the vertex

(x_n, y_n) is assumed to be the same as (x_0, y_0). Note that if the points are numbered in clockwise order the area A, computed as above, will have a negative sign; but the centroid coordinates will be correct even in this case.

Centroid of Cone or Pyramid

The centroid of a cone or pyramid is located on the line segment that connects the apex to the centroid of the base.

For a solid cone or pyramid, the centroid is 1/4 the distance from the base to the apex. For a cone or pyramid that is just a shell (hollow) with no base, the centroid is 1/3 the distance from the base plane to the apex.

Construction

In the fields of architecture and civil engineering, construction is a process that consists of the building or assembling of infrastructure. Far from being a single activity, large scale construction is a feat of human multitasking. Normally, the job is managed by a project manager, and supervised by a construction manager, design engineer, construction engineer or project architect.

For the successful execution of a project, effective planning is essential. Involved with the design and execution of the infrastructure in question must consider the environmental impact of the job, the successful scheduling, budgeting, construction site safety, availability of building materials, logistics, inconvenience to the public caused by construction delays and bidding, etc.

Types of Construction Projects

In general, there are three types of construction:

1. Building construction
2. Heavy / civil construction
3. Industrial construction.

Each type of construction project requires a unique team to plan, design, construct and maintain the project.

Building Construction: Building construction is the process of adding structure to real property. The vast majority of building construction projects are small renovations, such as addition of a room, or renovation of a bathroom. Often, the owner of the property acts as labourer, paymaster, and design team for the entire project. However, all building construction projects include some elements in common - design, financial, estimating and legal considerations. Many projects of varying sizes reach undesirable end results, such as structural collapse, cost overruns, and/or litigation reason, those with experience in the field make detailed plans and maintain careful oversight during the project to ensure a positive outcome.

Commercial building construction is procured privately or publicly utilising various delivery methodologies, including cost estimating, hard bid, negotiated price, traditional, management contracting, construction management-at-risk, design & build and design-build bridging. Residential construction practices, technologies, and resources must conform to local building authority regulations and codes of practice. Materials readily available in the area generally dictate the construction materials used (e.g. brick versus stone, versus timber). Cost of construction on a per square metre (or per square foot) basis for houses can vary dramatically based on site conditions, local regulations, economies of scale (custom designed homes are always more expensive to build) and the availability of skilled tradespeople. As residential (as well as all other types of construction) can generate a lot of waste, careful planning again is needed here.

The most popular method of residential construction in the United States is wood framed construction. As efficiency codes have come into effect in recent years, new construction technologies and methods have emerged. University Construction Management departments are on the cutting edge of the newest methods of construction intended to improve efficiency, performance and reduce construction waste.

Industrial Construction: Industrial construction, though a relatively small part of the entire construction industry, is a very important component. Owners of these projects are usually large, for-profit, industrial corporations. These corporations can be found in such industries as Infrastructure, Power Transmission & Distribution, metallurgical and material handling, medicine, petroleum, chemical, power generation, manufacturing etc. Processes in these industries require highly specialised expertise in planning, cost estimating, design, and construction. As in building and heavy/highway construction, this type of construction requires a team of individuals to ensure a successful project often undertaken by big construction companies.

Construction Processes

Design Team

In the modern industrialised world, construction usually involves the translation of designs into reality. A formal design team may be assembled to plan the physical proceedings, and to integrate those proceedings with the other parts. The design usually consists of drawings and specifications, usually prepared by a design team including surveyors, civil engineers, cost engineers (or quantity surveyors), mechanical engineers, electrical engineers, structural engineers, fire protection engineers, planning consultants, architectural consultants, and archaeological consultants. The design team is most commonly employed by (i.e. in contract with) the property owner. Under this system, once the design is completed by the design team, a number of construction companies or construction management companies may then be asked to make a bid for the work, either based directly on the design, or on the basis of drawings and a bill of quantities provided by a quantity surveyor. Following evaluation of bids, the owner will typically award a contract to the most cost efficient bidder.

The modern trend in design is toward integration of previously separated specialities, especially among large firms. In the past, architects, interior designers, engineers, developers, construction managers, and general contractors were more likely to be entirely separate companies, even in the larger firms. Presently, a firm that is nominally an "architecture" or "construction management" firm may have experts from all related fields as employees, or to have an associated company that provides each necessary skill. Thus, each such firm may offer itself as "one-stop shopping" for a construction project, from beginning to end. This is designated as a "design Build" contract where the contractor is given a performance specification and must undertake the project from design to construction, while adhering to the performance specifications.

Several project structures can assist the owner in this integration, including design-build, partnering and construction management. In general, each of these project structures allows the owner to integrate the services of architects, interior designers, engineers and constructors throughout design and construction. In response, many companies are growing beyond traditional offerings of design or construction services alone and are placing more emphasis on establishing relationships with other necessary participants through the design-build process.

The increasing complexity of construction projects creates the need for design professionals trained in all phases of the project's life-cycle and develop an appreciation of the building as an advanced technological system requiring close integration of many subsystems and their individual components, including sustainability. Building engineering is an emerging discipline that attempts to meet this new challenge.

Financial Advisors

Construction projects can suffer from preventable financial problems. Underbids ask for too little money to complete the project. Cash flow problems exist when the present amount of funding cannot cover the current costs for labour and materials, and because they are a matter of having sufficient funds at a specific time, can arise even when the overall total is enough. Fraud is a problem in many fields, but is notoriously prevalent in the construction field. Financial planning for the project is intended to ensure that a solid plan with adequate safeguards and contingency plans are in place before the project is started and is required to ensure that the plan is properly executed over the life of the project.

Mortgage bankers, accountants, and cost engineers are likely participants in creating an overall plan for the financial management of the building construction project. The presence of the mortgage banker is highly likely, even in relatively small projects since the owner's equity in the property is the most obvious source of funding for a building project. Accountants act to study the expected monetary flow over the life of the project and to monitor the payouts throughout the process. Cost engineers and estimators apply expertise to relate the work and materials involved to a proper valuation. Cost overruns with government projects have occurred when the contractor was able to identify change orders or changes in the project resulting in large increases in cost, which are not subject to competition by other firm as they have already been eliminated from consideration after the initial bid.

Large projects can involve highly complex financial plans and often start with a conceptual estimate performed by a consulting estimator. As portions of a project are completed, they may be sold, supplanting one lender or owner for another, while the logistical requirements of having the right trades and materials available for each stage of the building construction project carries forward. In many English-speaking countries, but not the United States, projects typically use quantity surveyors. In the United States construction estimating professionals like members of the American Society of Professional Estimators of the Consulting Estimators FORUM are typically employed.

Chapter 9

Multiple Force Members

The term, *three-force member*, is often incorrectly applied to members with more than three forces. Any member which is subjected to more than three single concentrated loads should be identified as a Multiple Force Member. The "extra" forces acting upon a multiple force member can be replaced until the total number of forces is reduced to three; the resultant force of the loading and the two reactions. This means that distributed loads must always be replaced so that only one resultant force is applied to the member. Then, all of the other forces must be replaced so that there is only one resultant representing the external loading. The three non-parallel forces must be concurrent so that the three-force principle then applies.

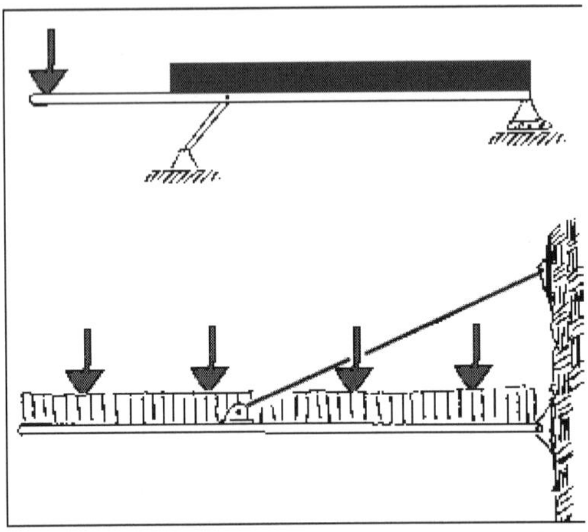

The loading of the first image would be resolved in the following manner. The resultant of the distributed load would be determined. This would act at the midpoint of the width of the load. The resultant of this force and the concentrated load acting at the cantilevered end of the beam would be combined into one load by summing moments about the right-hand support. This system would then be reduced from multiple force member to a three force member.

The second system could represent an entrance canopy to a hotel that also has four large spot lights sitting upon it. These six loads must also be combined into one load resultant before the member can be considered a three force member.

Primary Loads

The loading of a building structure can take on a wide variety of forms. In many cases the exact loading will not fit neatly into a specific category. Yet, loads can usually be considered to be Primary or Secondary. Primary loads are discussed below. Secondary loads are those loads due to temperature changes, construction eccentricities, shrinkage of structural materials, settlement of foundations, or other such loads.

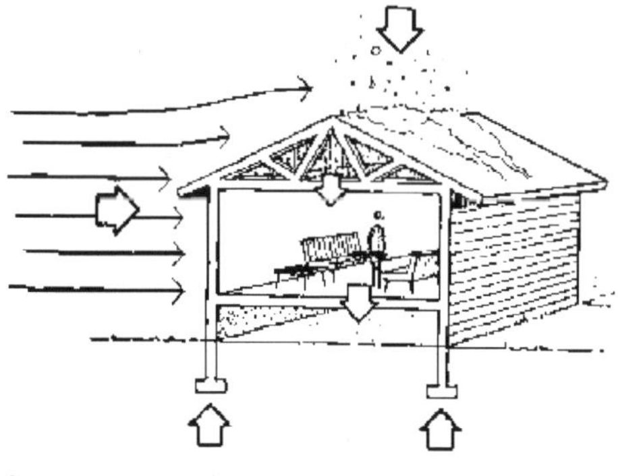

Despite the fact that each and every load and loading combination should be considered in order to reduce the chance of structural failure, the determination of the loading remains a statistical exercise. Each and every load cannot be foreseen; thus, it is critical to determine the worst case that is reasonable to assume to act upon the structure. The sources of primary loading include the materials from which the structure was built, the occupants, their furniture, and various weather conditions, as well as unique loading conditions experienced during construction, extreme weather and natural catastrophes. Primary Loads are divided into two broad categories according to the way in which they act upon the structure or structural element. These are DEAD LOADS and LIVE LOADS. When considering the possible combinations of these two categories of loading, the odds of certain loads occurring simultaneously are assumed to be null. One such combination would be heavy snow, a typhoon, a raging fire and an earthquake. It is possible that two of the first three could occur as the earth quakes, but not that all four would be present at the same time. Thus, one need only consider reasonable loading combinations.

Dead Loads

Dead Loads are those loads which are considered to act permanently; they are "dead," stationary, and unable to be removed. The self-weight of the structural members normally provides the largest portion of the dead load of a building. This will clearly vary with the actual materials chosen. Permanent non-structural elements such as roofing, concrete, flooring, pipes, ducts, interior partition walls, Environmental Control Systems machinery, elevator machinery and all other construction systems within a building must also be included in the calculation of the total dead load. These loads are represented by the red arrow in the illustration.

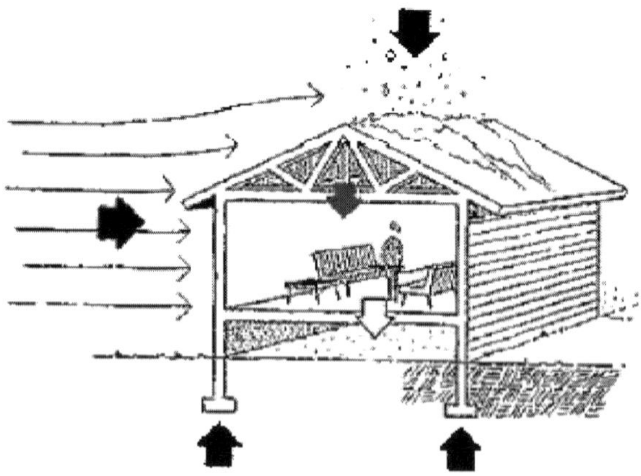

The magnitude of the dead loads of a building can usually be determined with only a 5% margin of error. Properties of building materials are often tabulated and published by authors of textbooks and by the manufacturers. It is always very important to be sure to keep up-to-date on the changes in building materials. Properties of construction materials often vary due to the rapidly changing marketplace. The load due to these materials is often expressed as kN/m^3 or lbs/ft^3. These are normally converted to load/area or load/length for further consideration.

Live Loads

Live Loads are not permanent and can change in magnitude. They include items found within a building such as furniture, pianos, safes, people, books, cars, computers, machinery, or stored materials, as well as environmental effects such as loads due to the sun, earth or weather. Wind and earthquakes loads are put into the special category of lateral live loads due to the severity of their action upon a building and their potential to cause failure.

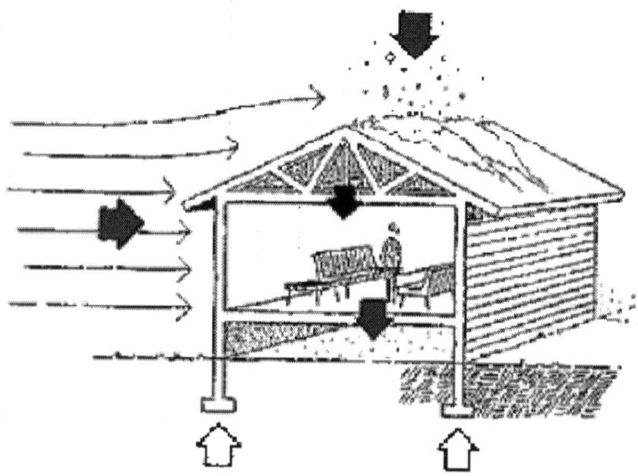

Most buildings have a working life that extends beyond the initially perceived use of the building. Thus, it is almost impossible to predict all of the potential uses that any given structure will experience before it is demolished. If, and when, a building is to be used for a purpose other than its original design, the capacity of the building for its new use must be determined. Since the body of knowledge about the behaviour of buildings is always increasing, a building which might have been designed according to the latest loading information in one year, might not satisfy the requirements a few years later. This has especially been the case with the effects of earthquake loading.

The magnitudes of live loads are difficult to determine with the same degree of accuracy that is possible with dead loads. The probable maximum value of live loads has been determined by research and is included in national building codes. These are usually a minimum design load per area. Building codes also provide for load reductions under certain conditions. As an example, full live loads will not occur on every floor of a multi-story building at the same time. Therefore, the design live load for some of the columns and the foundation can be reduced. Building codes around the world do not concur on the magnitude of the appropriate design live load values. It is critical that the designer take the time to determine the values set down in the local building codes. These are legal documents and MUST be followed.

Lateral Loads

Most lateral loads are live loads whose main component is a horizontal force acting on the structure. Typical lateral loads would be a wind load against a facade, an earthquake, the earth pressure against a beach front retaining wall or the earth pressure against a basement wall. Most lateral loads vary in intensity depending on the building's geographic location, structural materials, height and shape. The dynamic effects of wind and earthquake loads are usually analysed as an equivalent static load in most small and moderate-sized buildings. Others must utilise the iterative potential of the computer. The design wind and earthquake loads on a building are substantially more complex than the following brief discussion and simple examples would indicate. The Uniform Building Code describes the design wind load determination in more detail for the various parts of the United States.

Wind Loads

The most common lateral load is a wind load. The Eiffel Tower is one example of a building which has a structure that was designed to resist a high wind load. Wind against a building builds up a positive pressure on the windward side and a negative pressure (or suction) on the leeward side. Depending upon the shape of the structure it may also cause a negative pressure on the side walls or even the roof. The pressure on the walls and roof is not uniform, but varies across the surface. Winds can apply loads to structures from unexpected directions. Thus, a designer must be well aware of the dangers implied by this lateral load. The magnitude of the pressure that acts upon the surfaces is proportional to the square of the wind speed.

Wind loads vary around the world. Meteorological data collected by national weather services are one of the most reliable sources of wind data. Factors that effect the wind load include the geographic location, elevation, degree of exposure, relationship to nearby structures, building height and size, direction of prevailing winds, velocity of prevailing winds and positive or negative pressures due to architectural design features (atriums, entrances, or other openings). All of these factors are taken into account when the lateral loads on the facades are calculated. It is often necessary to examine more than one wind load case. For this course, it will be assumed that wind loads, as well as the pressure they develop upon wall and roof elements, are static and uniform. They actually not only pound a structure with a constantly oscillating force, but also increase as a building increases in height. The loading of a tower can be very roughly approximated by an evenly distributed load. It is a vertical cantilever. The applet below allows you to investigate the variables which influence the structural behaviour of a tall, thin tower. It does not represent actual methods of calculating the total wind force on a tall building. It is intended to demonstrate the interaction between the variables of the equations which govern the structural behaviour.

Earthquake Loads

Earthquake loads are another lateral live load. They are very complex, uncertain, and potentially more damaging than wind loads. It is quite fortunate that they do not occur frequently. The earthquake creates ground movements that can be categorised as a "shake," "rattle," and a "roll." Every structure in an earthquake zone must be able to withstand all three of these loadings of different intensities. Although the ground under a structure may shift in any direction, only the horizontal components of this movement are usually considered critical in a structural analysis. It is assumed that a load-bearing structure which supports properly calculated design loads for vertical dead and live loads are adequate for the vertical component of the earthquake. The "static equivalent load" method is used to design most small and moderate-sized buildings.

The lateral load resisting systems for earthquake loads are similar to those for wind loads. Both are designed as if they are horizontally applied to the structural system. The wind load is considered to be more of a constant force while the earthquake load is almost instantaneous. The wind load is an external force, the magnitude of which depends upon the height of the building, the velocity of the wind and the amount of surface area that the wind "attacks." The magnitude earthquake load depends up the mass of the structure, the stiffness of the structural system and the acceleration of the surface of the earch. It can be seen that the application of these two types of loads is very different. This movie is a representation of the movement of a free standing water tower in an earthquake. It can be seen that the as the ground moves, the initial tendency is for the water tower to remain in place. The shifting of the ground is so rapid that the tower cannot "keep up."

After a moment, the tower moves to catch up with the movement of the ground. The movement is actually an acceleration. From Newtonian Physics, it is know that an applied force=mass x acceleration. Thus, the force which is applied to the water tower depends upon the mass of the tower and the acceleration of the earth's surface. The force in this last diagram may be thought of as the "equivalent static load" for which the structure would be designed. This idealised situation demonstrates a concept; it requires modification for actual buildings. These modifications account for building location, importance, soil type, and type of construction. This movement can also be seen in the following movie of lateral earth movement. Note how the mass slowly reacts to the movement of the earth. Eventually, the bending strength of the stem of the tower would be exceeded and it will fail.

It remains very difficult to imagine the destruction which can be wrought by an earthquake. The lessons learned from the Los Angeles Earthquake of 1994 helped structural designers change design strategies. The following illustrations are snapshots from LA:

Fluid and Earth Pressure Loads

Liquids produce horizontal loads in many structures. The horizontal pressure of a liquid increases linearly with depth and is proportional to the density of the liquid. This is similar for earth pressures. These last are a bit more complex in that the load due to earth pressure varies with its depth, any surcharge, the type of soil and its moisture content. The design live load for this soil pressure must not be less than that which would be caused by a fluid weighing 30 pcf.

Load Distribution

Gravity Load Distribution

Architects must consider not only what loading conditions might effect a structure, but also how those loads are transferred from their point of origin through the structure to the ground. Architectural design decisions concerning the structural framing system will dictate this flow of forces. The most direct path from the point of loading to the ground will create the most efficient structure. The continuity of this line of transfer is essential. However, this direct path is not always possible. As a matter of fact, due to the nature of live loads this path is constantly changing!

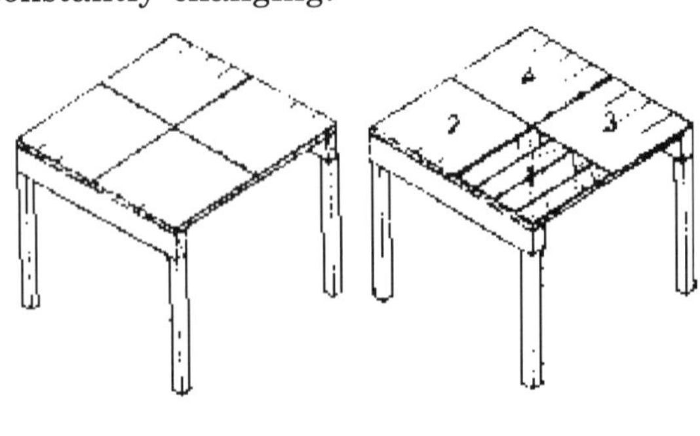

The load at the midspan of the joist of the frame is illustrated as a large arrow pointing down. This load is split into two parts, with a proportional part going to each end of the beam. This in turn becomes a point load at the midspan of the smaller beams. They transfer the load to the earth. The load on each column in this case is only one quarter of the initial load. This is one method of distributing a load.

The second drawing illustrates a wooden floor of thick planks that span between two beams. Each of the four squares of the flat surfaces is the tributary area of the beam below it. A tributary area of a structural element (such as a joist, beam, column, or wall) is the area that contributes to the loading of that specific element. In this case, the plans transfer their loads and pass them on to the two beams. These beams can only transfer their loads equally to the two columns which support them. Thus, the tributary area can be drawn by determining the supporting characteristics of the horizontally spanning members.

The loads are gathered by each structural element and passed on to that element's supporting elements. These supports in turn pass their loads to the next supporting element until the original load has been transferred all the way to the earth.

The load distribution pattern and the explicit summation of the loads can have a direct effect upon the size of the elements. Load distribution often causes unequal loading of the vertical supporting members. This may or may not be indicated by the designed form of that element.

Wind Load Distribution

The essence of wind load distribution through a building from one building component to another, and how the building resists this load can be investigated by assuming a uniformly distributed wind load acting on any one face of the building at any one time. Normally a wind load design requires a separate analysis of wind from two perpendicular directions, such as wind from the north or south and then from the east or west. Consider a one story, flat roofed, rectangular, wood framed building without a parapet. The wall receiving the direct pressure from the wind distributes the top half of its horizontal wind load to the roof and the bottom half to the foundation; it acts as a vertical beam with the roof and foundation acting as simple horizontal supports. The part of the load going to the foundation is distributed from the foundation to the ground. The portion of the load going to the roof tends to cause the roof to move laterally; this lateral movement is resisted by the end walls. The movement of the end walls is prevented by their connection to the foundation.

Since winds of 80 to 110 mph will create pressure on the side walls of buildings in the range of 15 to 30 psf, design forces of the wind will be considered between 20 or 25 psf (static load) in this course.

Resolving Distributed Loads

Many structural members are subjected to distributed loads rather than concentrated loads. However, when solving for their reactions, it is often convenient to replace the distributed loads with an equivalent concentrated load(s).

The most common type of distributed load is the uniform load, in which a load of constant magnitude is applied along a length of a beam or an area of a surface. The resultant of these distributed loads has the same magnitude as the area of the distributed load and acts through the centre of gravity, or the centroid, of the load area. This point is located at the intersection of the diagonals of a rectangular, or uniform, load area.

Not all distributed loads are of a constant value. Some examples of loading conditions that may cause triangular loads are:

- the load on a lintel in a masonry wall due to the arching of the masonry;
- the load on a vertical surface below grade due to soil or water pressure;
- the load on the vertical surface of a structure due to wind loading.

These triangular loads can be replaced by a single concentrated load which acts through the centroid of the distributed load and acting in the same direction. The centroid of a triangular load area is found at the intersection of its medians which always equals 1/3 of the height above any of its bases.

Triangular loads can be caused by either soil or water pressure. Note that it constantly increases with depth! The magnitude of the greatest load that is applied is simply the unit weight of the material multiplied by the deepest point in question.

There are cases where a uniformly distributed load combines with a triangularly distributed load to create a trapezoidal load. This loading condition is most easily solved by separating the triangular and uniform loads and calculating separate resultants for each. These resultants can then be treated as a series of concentrated loads or transformed into a single resultant for the system.

Strength of Materials

A structural system is not only effected by external conditions, but also by the properties and behaviour of the materials which comprise it. These also determine the nature of the system's reaction(s) to external forces. The study of Strength of Materials is concerned specifically with the following issues:

1. The internal forces of a member caused by the external forces acting on that member or system.
2. The changes in dimension of a member caused by these forces.
3. The physical properties of the material in the member.

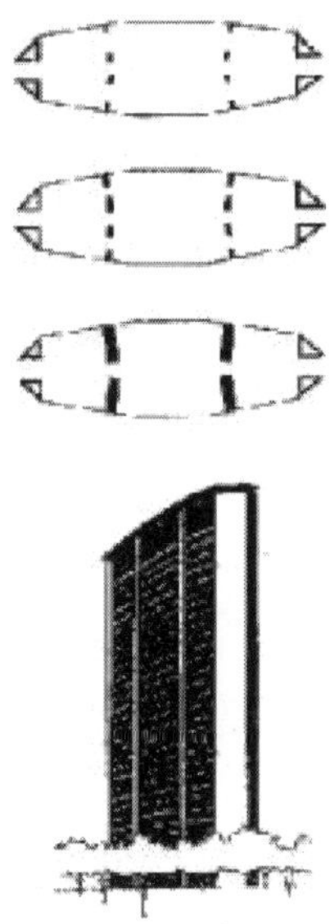

Statics is the study of the behaviour of rigid bodies at rest as they are acted upon by external forces. Although most of these bodies are not absolutely rigid, the assumption of rigidity is valid for the purpose of determining the reactions of the system. Actually, every material will deform under a load. Even a concrete slab deforms microscopically when a person walks on it. Some deformations in a structure can be detrimental to the overall system's performance, while others might only be an issue of comfort. The recognition of the relative importance of these deformations will be an important part of the study of structures. External loads on a structural system create resisting forces within all of the members that form the load path from the point of the application of the load to the ground beneath the foundation. This internal resistance exists within every member and joint included in the load path and are known simply as the internal forces acting on a member.

Some of these forces have already been examined: the connection between a beam and its support, and the connection of a two-force member to a three-force member in a pin-connected frame. The internal forces within a beam were demonstrated by cutting a beam and drawing a FBD. These internal forces were required at the cut section to put the beam back into equilibrium. The forces and moments that were examined were applied externally to the end of that cut section; they were exactly equal to the internal forces and moments. The distribution of the internal forces on the cross-sectional area of a member may or may not be uniformly distributed; it is dependent on the loading condition, the type of member, and how it is supported.

Stress and Stress Prisms

The internal forces of each member of a structural system are distributed in a particular way across that member's cross-sectional area. Stress is a measure of the intensity of this force on a single unit of area. Common units of measure of stress are KN/m^2, N/mm^2, #/in^2, #/ft^2, or kips/in^2. The distribution of the stress may be constant across a cross-sectional area, or it may be variable. It can vary due to the loading conditions, the material or geometry of the structural member.

Structural members are subjected to a variety of types of stress. They can be divided into primary and secondary stresses. Primary stresses are Compression, Tension, Torsion and Shear. Secondary stresses are the result of deformations of the structural system, or member, which cause additional loads within each member to occur (For example, a beam expanding in the summer sun pushing on a fixed column). These stresses will be discussed in more detail in a later lecture. Initially, the focus will be on simple tension and compression stresses for members loaded symmetrically along an axis, which causes uniformly distributed internal stress.

The Pirelli Office Block, designed by Ponti and engineered by Nervi, is a wonderful example of a structure in which the columns express not only the relationship between the moment and the building, but of the relationship between direct stress and size as well. The stress in the column was intended to remain as constant as possible.

Stress-Strain Curves

The relationship between the stress and strain that a material displays is known as a Stress-Strain curve. It is unique for each material and is found by recording the amount of deformation (strain) at distinct intervals of tensile or compressive loading. These curves reveal many of the properties of a material (including data to establish the Modulus of Elasticity, E). What does a comparison of the curves for mild steel, cast iron and concrete illustrate about their respective properties?

It can be seen that the concrete curve is almost a straight line. There is an abrupt end to the curve. This, and the fact that it is a very steep line, indicate that it is a brittle material. The curve for cast iron has a slight curve to it. It is also a brittle material. Both of these materials will fail with little warning once their limits are surpassed. Notice that the curve for mild steel seems to have a long gently curving "tail". This indicates a behaviour that is distinctly different than either concrete or cast iron. The graph shows that after a certain point mild steel will continue to strain (in the case of tension, to stretch) as the stress (the loading) remains more or less constant. The steel will actually stretch like taffy. This is a material property which indicates a high ductility. There are a number of significant points on a stress-strain curve that help one understand and predict the way every building material will behave. An example plot of a test on two grades of steel is illustrated above. If one begins at the origin and follows the graph a number of points are indicated. Point A is known as the proportional limit. Up to this point the relationship between stress and strain is exactly proportional. The number which describes the relationship between the two is the Modulus of Elasticity. This is discussed in more detail in the next lecture.

Strain increases faster than stress at all points on the curve beyond point A. Up to this point, any steel specimens that is loaded and unloaded would return to its original length. This is known as elastic behaviour. Point B is the point after which any continued stress results in permanent, or inelastic, deformation. Thus, point B is known as the elastic limit. Since the stress resistance of the material decreases after the peak of the curve, this is also known as the yield point. The line between points C and D indicates the behaviour of the steel specimen if it experienced continued loading to stress indicated as point C. Notice that the dashed line is parallel to the elastic zone of the curve (between the origin and point A). When the specimen is unloaded the magnitude of the inelastic deformation would be determined (in this case 0.0725 inches /inch). If the same specimen was to be loaded again, the stress-strain plot would climb back up the line from D to C and continue along the initial curve. Point E indicates the location of the value of the ultimate stress. Note that this is quite different from the yield stress. The yield stress and ultimate stress are the two values that are most often used to determine the allowable loads for building materials and should never be confused.

A material is considered to have completely failed once it reaches the ultimate stress. The point of rupture, or the actual tearing of the material, does not occur until point F. It is interesting to note the curve that indicates the actual stress experienced by the specimen. This curve is different from the apparent stress since the cross sectional area is actually decreasing. There is quite a bit to be learned from both the study of the ideal and actual behaviour of all building materials. Changes in that body of knowledge have had large impacts on the way in which building structures are designed.

The earliest methods of design limited the stresses that a structure would be "allowed" to experience. Thus, the method of design was known as the Allowable Stress Method. Recognition of the additional strength potential of most materials resulted in the Ultimate Stress Method of design. Contemporary thought centres on the limitation of the various service conditions of the structure at hand. This is known as the Limit States Design method. In the end, it is the authors opinion that the actual method of design is less important than the legal bodies would like us to believe. Human factors in the construction process SHOULD prevent a good designer from pushing too hard against the envelope of safety.

Elasticity and Deformation

An important property of many structural materials is their ability to regain their original shape after a load is removed. These materials are called *elastic*. Steel, glass and rubber are elastic; putty or modelling clay are not elastic. Each of these materials is elastic to varying degrees; steel and glass are both more elastic than rubber. The degree of elasticity, or "stiffness" of a material is called its Modulus of Elasticity (E). Given the modulus of elasticity, possible deformations can be calculated for any material and loading.

All materials are assumed to be elastic except when indicated otherwise in this course. Robert Hooke (1635 - 1703), a great English scientist, experimented with springs, clocks and watches. During his investigation of the spring he discovered that in elastic materials, stress and strain are proportional. He first presented this in a lecture in 1678 and it is known today simply as Hooke's Law.

Hooke's Law applies as long as the material stress does not pass a certain point known as its proportional limit. This is the point at which the physical properties of the material actually change. Any time an elastic material is loaded between zero and the proportional limit, the stress and strain are directly proportional and if the load is released the material will regain its initial dimensions. If the stress is doubled the strain is doubled; if the stress is tripled the deformation is three times as great, etc.

The English physician and physicist Thomas Young (1773 - 1829) noted that if stress is proportional to strain, then for any given material, stress divided by strain would be a constant. This constant is known today as Young's Modulus or the Modulus of Elasticity.

The Modulus of Elasticity is represented by E = Stress / Strain.

This relationship is found as the slope of the curve of the stress-strain curve from initial loading to the proportional limit. A higher value of the modulus indicates a more brittle material (i.e. glass, ceramics). A very low value represents a ductile material (i.e. rubber).

The values of the modulus of elasticity for structural materials have been determined by tests and are readily available in references such as the AISC manual. Some of the more common values are:

- Steel: E= 29,000 KSI (sometimes rounded to 30,000 KSI)
- Aluminum: E= 10,000 KSI
- Wood: E= 1,000 - 2,000 KSI (usual range)
- Concrete E= 2,500 - 5,000 KSI (usual range).

It is often necessary to be able to determine the deformation of a structural member once the loads and physical properties of the structural member are known. This is simply derived and is developed from the stress/strain relationships that have already been established.

Secondary Loads

Some types of loads on buildings are due to deformation. The deformation may be due to earth sliding or settling unevenly under the footings. This type of load can be avoided by properly designed footings on stable soil. The deformation of a building may be caused by temperature change. To avoid excessive stresses, construction must provide for a certain amount of thermal expansion and contraction.

Thermal Stresses

The size of a body will change as the ambient temperature fluctuates, expanding as it rises and contracting as it falls. For example, cracks in an asphalt highway which were there all winter, disappear during the summer. The asphalt closes the cracks as it expands in the summer heat but they will reappear in the cold of winter as the material shrinks. This repetitive cycle of expansion and contraction results in significant temperature stresses in construction material.

Within the ordinary climatic temperature range experienced by buildings, the change in dimension has conveniently been found to be approximately proportional to the change in temperature. The amount of dimensional change that will take place is determined by a factor known as the coefficient of thermal expansion (alpha). This describes the change in length (or width) of a member per unit length of the member. The values below describe some of the coefficients of thermal expansion for some common materials.

***Table** : Coefficients of Thermal Expansion Per degree F*

Wood	0.000 0030
Glass	0.000 0044
Concrete	0.000 0060
Cast Iron	0.000 0061
Steel	0.000 0065
Wrought Iron	0.000 0067
Copper	0.000 0093
Bronze	0.000 0100
Brass	0.000 0104
Aluminum	0.000 0128

These values describe the change in length per unit of length for one degree F temperature change. Some tables list the coefficients for a 100 F temperature change. Similar tables may be found for the temperature change based on Centigrade rather than Fahrenheit. These are not only used to determine the values for expansion, but for contraction or shortening as well.

The equation used in determining the deformation is as follows:

Delta L = (alpha) (delta T)(L)

Where:

delta L= total change in length, inches (total deformation)

alpha = coefficient of thermal expansion

delta T = temperature change

L = length in inches

Note that the equation does not include a variable for the cross-sectional area. Every construction material will expand and contract with changes in temperature; regardless of the cross-sectional area.

Built-up Sections

It is often advantageous to combine a number of smaller members in order to create a beam or column of greater strength. The moment of inertia of such a built-up section is found by adding the moments of inertia of the component parts. This can be done< B> if and only if the moments of inertia of each component area are taken about a common axis and if and only if the resulting section acts as one unit.

Under No Other Condition Can They Be Added!

Two examples of built-up sections are seen below. In each case the components of the whole have a common axis and act as one unit.

Transfer Formula

There are many built-up sections in which the component parts are not symmetrically distributed about the centroidal axis. The easiest way to determine the moment of inertia of such a section is to find the moment of inertia of the component parts about their own centroidal axis and then apply the transfer formula. The transfer formula transfers the moment of inertia of a section or area from its own centroidal axis to another parallel axis. It is known from calculus to be:

$I_x = I_c + (A)d^2$

Where:

I_x = moment of inertia about axis x-x (in^4)

I_c = moment of inertia about the centroidal axis c-c parallel to x-x (in^4)

A = area of the section (in^2)

d = perpendicular distance between the parallel axes x-x and c-c (in)

Finding the moment of inertia of an asymmetric built-up cross-section is simplified to the procedure shown diagrammatically below:

Columns

The column is one of the most familiar architectonic icons. Each of us can picture this very important space definer in our mind's eye.

It can delineate, punctuate, or separate spaces. A quick glance out most windows allows a view of at least one of them. Columns are vertical load-bearing elements which are normally loaded in compression. Time has brought many manifestations of this simple element; thick, thin, long, short, spindle shaped, squat, etc. And they have been made out of any material that has a minimal compressive strength.

The transition between the column and its loading member, usually a beam of some sort, has often been of great interest to builders. The structural/functional problem is quite simple - the forces collected by the beam must be transferred from the beam to the column - and yet the solutions demonstrated a great variation. The early Egyption, Greek and Roman civilizations turned this transition zone into an articulated flourish to draw attention to the capital. In almost every case it was broader at the top and tapered to the column shaft. This not only facilitated a smooth transition for the forces, but also eased the construction and even increased the stability of the buildings.

A column either crushes (a strength failure) or it buckles (a stability failure). Both modes of failure must be considered for every column. The exact mode of failure is greatly dependent upon the way in which the column's cross-sectional area is distributed with respect to its centroid. The following simple concept must be satisfied at all times:

Stress Due to Loading < Resistance Potential of the Column

This states that the stress within the column due to all of the applied loads must be less than the allowable stress of the material. This is a logical statement that is the essence of the structural analysis of a column. The actual failure mechanism could be due to a combination of two or more loads. These combinations must be carefully considered.

Column Failure Modes

Crushing: Relatively short columns are more apt to fail by the material crushing. Every building material material can withstand a distinct amount of compressive stress before it crushes. This value has been determined by laboratory tests and is known as the compressive strength of a material. This strength is dependent upon the internal structure of the material and/or its components. Steel has a very homogeneous, finely crystalled internal structure and has a relatively high compressive strength. Hard woods are fine-grained and have a higher compressive strength than soft-woods. Wood is unique in that it has two compressive strengths; one when loaded parallel to the grain and another when loaded perpendicular to the grain. Why is that? When a wood column crushes the fibres of the wood actually split apart. In every case, crushing is a strength failure and does not depend upon the shape of the section.

Buckling: Relatively slender columns are more apt to fail by buckling. A column is slender when it has a "small" cross-section compared to its effective length. Small is placed in quotes due to the fact that the important information about the cross section is both the actual size and more importantly, the shape of the cross-section. This is then compared to the effective length to determine whether or not the column is slender. If it is, this means that the column will probably fail in bending! As a column is loaded, it is likely to bend about the weak axis of the cross-section (the one with the lowest Moment of Inertia). A column buckles when it bends about an axis. This is a stability failure.

Radius of Gyration and Buckling

The radius of gyration (r) describes the way in which the area of a cross-section is distributed around its centroidal axis. If the area is concentrated far from the centroidal axis it will have a greater value of r and a greater resistance to buckling. A cross-section can have more than one radius of gyration and most sections have at least two. If this is the case, the section tends to buckle around the axis with the smallest value. The radius of gyration is defined as:

r = sqrt (I/A)

where

r = radius of gyration

I = moment of inertia

A = area of the cross section

All things being equal, a circular pipe is the most efficient column section to resist buckling. This is because it has an equal radius of gyration in all directions and it has the its area distributed as far away as possible from the centroid.

The steel columns shown below all have areas of 3-1/8 in^2. The safe loads for an 8 ft length are shown. The only difference between them is the way in which the cross-sectional area is distributed about the centroid.

Buckling

Buckling is very similar to bending. Thus, the shape of the cross-section is very important. The shape of the column also effects the way in which it will buckle. Imagine for a moment a single sheet of paper (A4 or 8.5 x 11). If one would try to simply stand it on edge it would be impossible unless the paper was folded. This simple act of folding the paper actually increases the cross-sectional moment of Inertia and thus the stiffness of the newly formed column.

The stiffness of another paper column could be further increased by taking the paper and taping the long edges together to create a tube. Now, the paper will be very stiff since the material of the paper is distributed evenly as far away from the neutral axis as possible.

The load at which a column will begin to buckle is known as the Critical Buckling Load (or critical load). A number of qualities of the column must be known in order to determine this value. The Swiss mathematician Leonard Euler (1707 - 1783) derived a formula in 1744 (known as the Euler Buckling Formula) to determine the load at which a perfect column will buckle.

It was a very important step in the history of technology and remains important for column design today. The equation is only accurate for columns which approach the perfect conditions for which he derived the equation.

$N_{cr} = ¼^{2} E\ I\ /\ (l_k^{2})$

In which the terms are defined as follows:

N_{cr} = Critical Buckling Load

E = Modulus of Elasticity

I = Moment of Inertia

l_k = Effective Buckling Length

This equation can be modified by dividing both sides by the area of the column so that the stress at which the column will buckle can be determined:

$f_{cr} = sigma_{cr} = N_{cr}/A = ¼^{2}\ E\ I\ /A(l_k^{2})$

now, knowing that $r^2 = I/A$, this equation becomes:

$= ¼^{2}\ E\ /(l_k/r)^2$

Lambda, or the slenderness ratio is a value with which one can gage the relative resistance of a column cross-section to buckling. Or, stated otherwise, the relative ease in which a column WILL buckle. It is defined as

lambda = l_k / r

Where l_k is the buckling length and r is the radius of gyration.

Thus, the critical buckling stress can be expressed as,

$$f_{cr} = ¼^2 \, E \, / \, (\text{lambda})^2$$

The buckling length of a column depends on its physical length and its end conditions. Euler discovered that if a column is hinged at both ends it will buckle in the form of a sine curve with the inflection points at the hinges. This would be the case in which the buckling length of a column is identical to its length. This is not the case if the ends of the column are both fixed. The determination of the buckling lengths for various column end conditions and frames is given below:

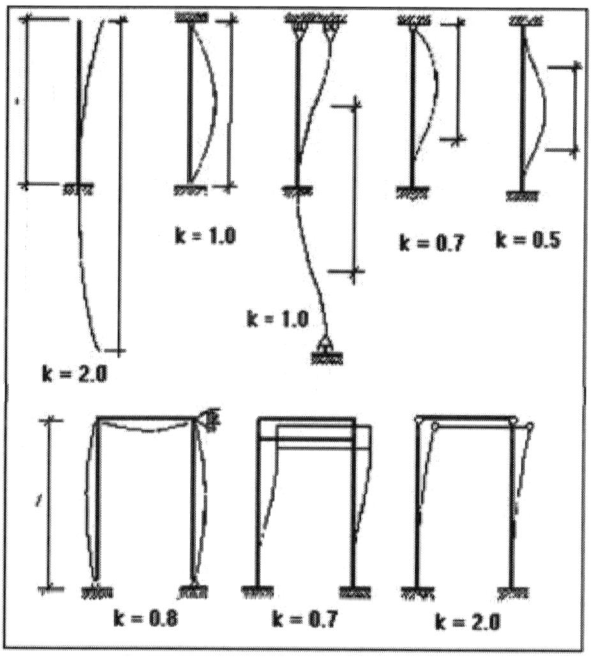

The magnitude of the internal forces is important to know in order to size a column. A small amount of tensile stress has little effect on a wood or steel column, but could cause problems in a concrete or masonry column.

Beams

A beam is a structural member which carries loads. These loads are most often perpendicular to its longitudinal axis, but they can be of any geometry. A beam supporting any load develops internal stresses to resist applied loads. These internal stresses are bending stresses, shearing stresses, and normal stresses.

Beam types are determined by method of support, not by method of loading. Below are three types of beams that will be investigated in this course:

1. Simple

2. Cantilever

3. Continuous

The first two types are statically determinate, meaning that the reactions, shears and moments can be found by the laws of statics alone. Continuous beams are statically indeterminate. The internal forces of these beams cannot be found using the laws of statics alone. Early structures were designed to be statically determinate because simple analytical methods for the accurate structural analysis of indeterminate structures were not developed until the first part of this century. A number of formulas have been derived to simplify analysis of indeterminate beams.

The three basic beam types can be combined to create larger beam systems. These complex systems can *inevitably* be distilled to the simple beam types for analysis. The beams shown immediately below are combinations of the first two beam types; these systems are all statically determinate.

Concentrated

Either a force or a moment can be applied as a concentrated load. Both are applied at a single point along the axis of a beam. These loads are shown as a "jump" in the shear or moment diagrams. The point of application for such a load is indicated in the diagram above. Note that this is NOT a hinge! It is a point of application. This could be point at which a railing is attached to a bridge, or a lamppost on the same.

Distributed

Distributed loads can be uniformly or non-uniformly distributed. Both types are commonly found on all kinds of structures. Distributed loads are shown as an angle or curve in the shear or moment diagram. A uniformly distributed load can evolve into a n on-uniformly distributed load (snow melting to ice at the edge of a roof), but are normally assumed to act as given. These loads are often replaced by a singular resultant force in order to simplify the structural analysis.

Beam Failure Modes

Structures fail in many ways; two categories of which have been described: stability failures and strength failures. Stability failures usually relate to structural systems, whereas strength failures relate to the members comprising a structure. Of the many ways which a beam can fail, three will be discussed in this course: bending, vertical shear, and horizontal shear.

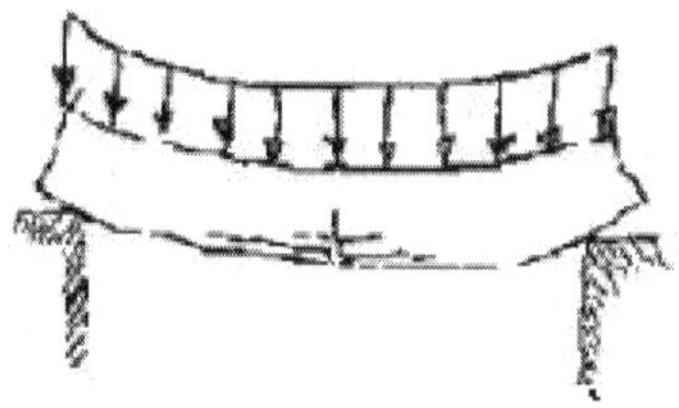

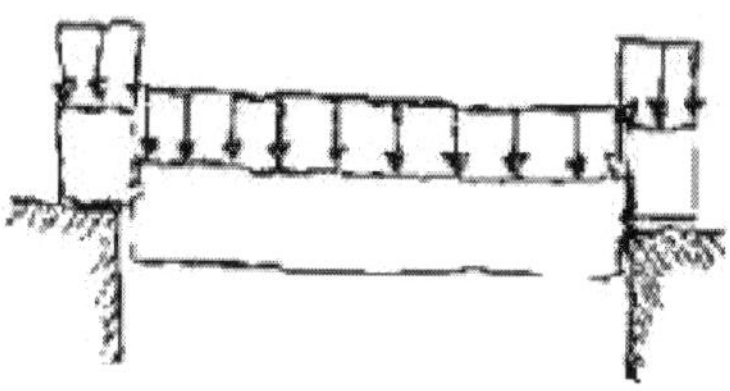

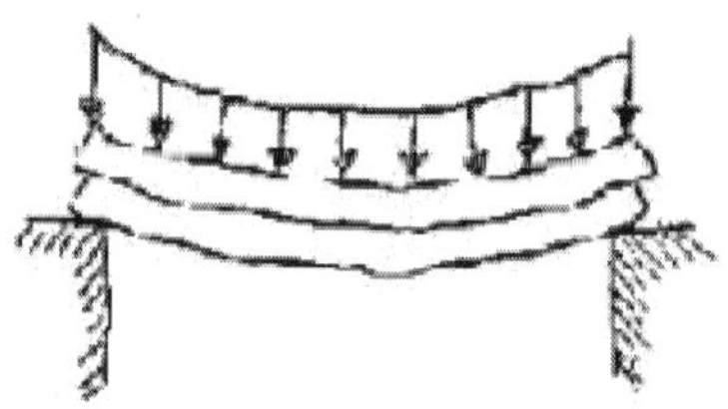

Bending is probably the most common type of failure. It is illustrated by the top figure in which the "fibres" along the bottom face of the beam are torn and those along the top face of the beam are crushed. Vertical Shear is an idealised mode of failure. There is a tendency for a short beam to fail in this manner. This is the way in which a pair of "shears" or scissors cuts a piece of paper. Horizontal Shear is the tendency for a material to separate parallel to the neutral axis as its "internal layers" try to slide past each other. It is a frequent mode of failure which should not be confused with checking in wood beams. Our discussion during the next few days will be limited to the stresses that are related to these modes of failure. Bending and horizontal shear are the types of failure most often encountered. Vertical shear is useful to us primarily because of what it reveals about horizontal shear. Excessive deflection can also be a cause of failure.

Internal Forces

Internal forces are generated within loaded structural elements. These forces are generated within every type of element; if they were not developed, the structure would fail. These are known as Shear, Moment, and Normal Forces. The normal force is found in columns and beams with an axial load. Shear and moment are found in beams and frames. Most of the elements that will be analysed in this course will be beams, including joists, purloins, girders, decking, planking, etc. Shear and moment are essential for developing an understanding of how structures behave. Considerable time will be spent on this subject because of its importance now and in the future. The capacity to draw shear and moment diagrams for any designed structure will add to the repertoire of the designer. All of the great Modern architects had this basic skill. In order to quantify the internal forces of any part of a beam a FBD must be drawn for *only* that specific part of the beam. If the entire structure is in equilibrium, this part, when isolated from the remainder as a Free Body, must also be in equilibrium. The action of the external forces on that part must be resisted by the internal forces acting at the cut section(s).

External Moment = Internal Moment or External Moment - Internal Moment = 0

Internal resisting forces and moments caused by the loading on a beam usually change along the length of the beam. Let us again examine the internal forces of a weightless beam by means of a FBD:

A beam carrying a load is clearly in equilibrium. If we cut the beam and apply the forces on the cut section which existed internally before the cut, the beam will still be in equilibrium. We can find these forces by applying equations of equilibrium on the ends of the cut sections (as in any FBDs). The solutions to the equations of equilibrium are the internal normal (N), shear (V) and moment (M) forces at that section. This is then applied to each and every point along the beam.

Shear (V) and Shear Diagrams

Shear (V) is the tendency for one part of a beam to slide past another part. The magnitude of the shear at any section is equal to the algebraic sum of loads and reactions acting perpendicular to that section.

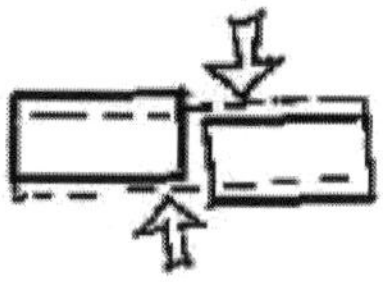

Sign Convention

If the tendency of the section to the left of the cut is to move upward, the shear is positive; if it has a tendency to move down, it is negative. For simplicity in obtaining the correct sign, one may say that it is equal to the algebraic sum of loads and reactions to the left of the cut. Note that the internal shear on the end of the FBD acts in the opposite direction of the algebraic sum of the loads and reactions in order to balance these forces (SV = 0).

A shear diagram is a graphic representation of the shear at every point along the length of a member.

To plot a shear diagram on a beam, the shear must be calculated at each point along the length of the beam. One way is to simply plot the shear as the algebraic sum of the loads and reactions acting perpendicular to the beam at the left side of each increment along the length of the beam. Positive values are shown above and negative values below a reference axis.

How to Draw a Shear Diagram

Start at the left end and plot the external shear values with regard to the following:

- The shear diagram is the graphic representation of the shear force at successive points along the beam. Forces acting upward are assumed positive and downward forces negative.
- The shear force (V) at any point is equal to the algebraic sum of the external loads and reactions, perpendicular to the beam, to the left of that point.
- Since the entire beam must be in equilibrium (sum of V = 0), the shear diagram must close to zero at the right end.
- Consider the loading for increments along the length of the beam in order to determine the shape of the curve.

If there is no change in the load along the incremental length under consideration, the shear curve is a straight horizontal line (or a curve of zero slope). The slope at any point is defined as the tangent to the curve at that point.

If a load exists, but does not change in magnitude over successive increments (uniformly distributed), the slope of the shear curve is constant and non-horizontal.

If a load exists, and increases in magnitude over successive increments, the slope of the shear curve is positive (approaches the vertical); if the magnitude decreases, the slope of the shear curve is negative (approaches the horizontal).

- Abrupt changes in loading cause abrupt changes in the slope of the shear curve. Concentrated loads produce vertical lines (a jump) in the shear curve.

Moment (M) and Moment Diagrams

The Moment (M) within a beam is a representation of the magnitude of the internal couple found within the beam at any given point. A moment diagram is a graphic representation of the moment at every point along the length of the member. To plot a moment diagram of a beam, one must determine the magnitude of the moment at each point along the length of the beam and then determine whether it is a positive or negative moment. Positive values are shown above and negative values below a reference axis. Sign convention is probably the most troublesome part of moment diagrams. In North America, if the moment tends to cause the beam to curve upward it is positive; if the moment tends to cause it to curve downward it is negative. This is the opposite of the conventions for most of the rest of the world. The sign of the moment diagram is not necessarily the same sign of the moment found in the FBD or in the moment equation when determining its magnitude. A simply supported beam with gravity load always has a *positive* moment. A cantilever with gravity load always has a *negative* moment. A continuous beam with an overhang subject to a gravity load experiences both *positive and negative* moments.

How to Draw a Moment Diagram

Start at the left end and plot the bending moment values due to external loads and reactions with regard to the following:

- The moment diagram is the graphic representation of the magnitude of the bending moment at successive points along the beam.

- The bending moment for the moment diagram (M) at any point equals the sum of moments of the forces to the left about that point.
- Since the entire beam is in equilibrium (sum of M= 0), the bending moment diagram must close to zero at the right side.
- Consider the values of successive incremental ordinates along the length of the beam in order to determine the shape of the curve:

If the magnitudes of successive shear ordinates are constant, the moment curve has a constant slope at that increment. If the magnitudes of successive shear ordinates increase, the slope of the moment curve is positive (it approaches the vertical). If the magnitudes of successive shear ordinates decreases, the slope of the moment curve is negative (it approaches the horizontal).

- Abrupt changes in the shear diagram will produce abrupt changes in the slope of the moment curve. Concentrated moments produce vertical lines in the moment curve.

Plotting V, M and Deflection

A comprehensive procedure for the construction of shear, moment, and deflection diagrams for statically determinate beams is given below. Each and every one of these diagrams are plotted from left to the right and positive values are shown above, and negative values below, a reference axis. The procedure is as follows:

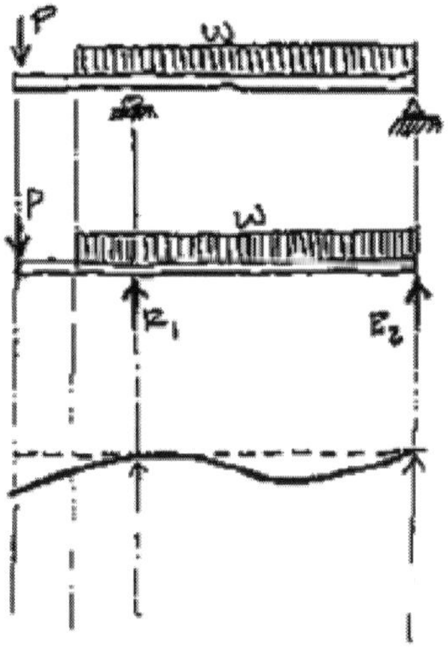

1. Draw a FBD of the entire structure and find the reactions of the beam.
2. Based on intuition and an educated guess, sketch an approximate deflection curve of the structure. This will help

to determine the location of inflection points and thus help to construct the diagrams.

3. Draw a SHEAR DIAGRAM. Start at the left end and plot the external shear values with regard to the following:
   - The shear diagram is the graphic representation of the shear force at successive points along the beam. Upward acting forces are assumed positive and downward forces negative.
   - The shear force (V) at any point is equal to the algebraic sum of the external loads and reactions to the left of that point.
   - Since the entire beam must be in equilibrium (the sum of V = 0), the shear diagram must close to zero at the right end.
   - Consider the loading for increments along the beam's length in order to determine the shape of the curve.

If there is no change in the load along the incremental length under consideration, the shear curve is a straight horizontal line (or a curve of zero slope). The slope at any point is defined as the tangent to the curve at that point.

If a load exists, but does not change in magnitude over successive increments (uniformly distributed), the slope of the shear curve is constant and non-horizontal.

If a load exists, and increases in magnitude over successive increments, the slope of the shear curve is positive (approaches the vertical); if the magnitude decreases, the slope of the shear curve is negative (approaches the horizontal).

   - Abrupt changes in loading cause abrupt changes in the slope of the shear curve. Concentrated loads produce vertical lines (a jump) in the shear curve.

4. Draw a MOMENT DIAGRAM. Start at the left end and plot the bending moment values due to the external loads and the reactions with regard to the following:
   - The moment diagram is the graphical representation of the magnitude of the bending moment at successive points along the beam.
   - The bending moment for the moment diagram (M) at any point equals the sum of moments of the forces to the left about that point.

- Since the entire beam is in equilibrium (Sum of M= 0), the bending moment diagram must close to zero at the right side.
- Consider the values of successive incremental ordinates along the length of the beam in order to determine the shape of the curve:

If the magnitudes of successive shear ordinates are constant, the moment curve has a constant slope at that increment. If the magnitudes of successive shear ordinates increase, the slope of the moment curve is positive (it approaches the vertical). If the magnitudes of successive shear ordinates decreases, the slope of the moment curve is negative (it approaches the horizontal).

- Abrupt changes in the shear diagram will produce changes in the shape of the moment curve. Concentrated moments produce vertical lines in the moment curve.

Deflection Diagram

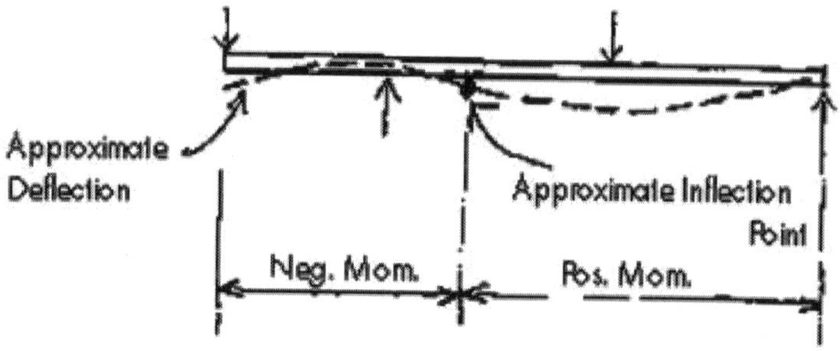

There is no set manner in which the deflection diagram should be drawn. There are, however, a few rules which can be followed:

- points of zero moment are inflection points (change in curvature direction) of the deflection curve.
- the maximum points of the moment diagram are the points at which the deflection diagram slope is zero (where the tangent of the curve is horizontal).

There are a few relationships (based on calculus) between these three diagrams that are helpful to construct them:

- The slope of the moment curve at any point is equal to the shear force at the same point.
- Points of zero shear are peaks of the moment curve, points of maximum positive or maximum negative moment.
- The change in moment value between any two points is equal to the area of the shear diagram between the same two points

(provided no other external moment is applied). This is known as the Shear Area Method.

- The shapes of the curves from shear to moment always progress in this order; straight horizontal line, straight sloped line, parabolic curve and cubic curve. Therefore, if a shear curve is a horizontal line, the moment curve is a sloped line; if th e shear curve is a sloped line the moment curve is a parabolic curved line; etc. This happens because of the mathematical relationship between loading, shear and moment diagrams.

Beam Stresses

Bending Stresses: A formula based on bending in a beam will be derived that can be used to:

- Select a beam that has adequate strength to carry a given load in bending.
- Determine the bending stresses in a given beam caused by a specific loading condition.
- Determine the greatest moment a given beam will resist (or the greatest load it will carry)

It will not be necessary to be able to derive the formula. However, it is essential to work through the derivation at least once in order to understand it and have confidence in how it is applied.

$$F = M/S \text{ or } S = M/F$$

In which,

S is the Section Modulus in in^3. This beam property can either be calculated or read from tables for most beams.

M is the moment in the beam in inch-pounds (or inch-kips). In the selection of a beam it is taken from the moment diagram.

F is the bending stress (psi or ksi) (Fb is usually used to denote the allowable bending stress; fb is used to indicate the actual bending stress).

Bending Theory

There are a number of assumptions that were made in order to develop the *Elastic Theory of Bending*. These are:

- The beam has a constant, prismatic cross-section and is constructed of a flexible, homogenous material that has the same Modulus of Elasticity in both tension and compression (shortens or elongates equally for same stress).

- The material is *linearly elastic*; the relationship between the stress and strain are directly proportional.
- The beam material is not stressed past its proportional limit.
- A plane section within the beam before bending remains a plane after bending.
- The neutral plane of a beam is a plane whose length is unchanged by the beam's deformation. This plane passes thought the centroid of the cross-section.

In order to visualize this, think of a black rubber beam with three lines drawn on its side. The dashed lines in the diagram below represent this beam, and the neutral axis and lines AB and CD are drawn parallel to their respective sides. Lines AB and C D are separated by some distance that is determined, but is not of consequence for the following discussion. These two lines were parallel before bending. As the beam bends, these lines remain perpendicular to the neutral axis.

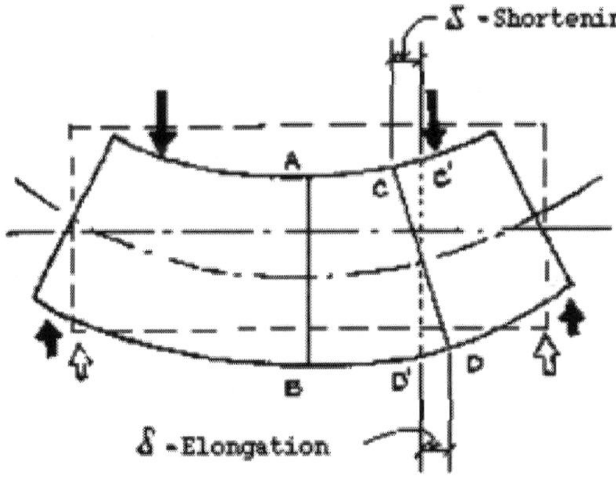

Thus, as the beam bends, developing a curve in the neutral plane that reflects the bending, the line CD does not remain parallel to line AD. There is a distinct shortening at the top face of the beam and elongation at the bottom face. CC' is equal to d (delta), or the shortening of the top fibre. Similarly, DD' is equal to (delta) or the elongation of the bottom fibre (tension). If these were measured, they would be found to be equal, but opposite. Knowing that the Modulus of Elasticity (E) describes a linear relationship between the strain and the amount of stress in a material and that this material is homogeneous and has a certain value for E, we can determine the stress. The magnitude of the strain at any point along C'D' can be found by using similar triangles. Therefore, the stress is also proportional to the distance from the neutral plane, as illustrated in the following diargram.

The loads and reactions acting on this beam segment cause a tendency for clockwise rotation (a clockwise moment). An internal moment counteracts this tendency so that the beam segment remains in equilibrium. Thus, the magnitude of the internal moment is exactly equal to the moment due to the external loads and reactions; but in the opposite direction. There is a state of equilibrium:

Internal Moment = External Moment

The illustration above shows the stress prisms developed due to the straining of the material. These prisms (compression and tension) can each be resolved into a single force that acts through the centroid of each prism. These two resulting forces (one is a compressive force (C) and the other a tensile force (T)) are found to have equal but opposite magnitudes. They are separated by a distance that is approximately 2/3 of the depth of the beam (because of their triangular geometry) and create an *internal couple(!)* which causes the internal moment resistance of the beam. The exact location of the resultant forces of the stress prisms depends upon the location of the centroid of the entire stress prism. The location of the centroid is in turn dependent upon the exact distribution of the stresses across the section which this is determined by the geometry of the cross-section.

General Case

If one accepts the assumptions given in the discussion of the bending theory above, the formula for determining the bending stress at any given fibre in a section can be derived as follows.

Given the beam as shown above loaded so that it has an internal bending moment. The stress at the extreme face (the distance C in this case) from the Neutral Axis (NA) is equal to Sigma or fb. The stress at any distance "y"; from the NA can be found using similar triangles. Thus,

$$fy = (fb/c)(y)$$

If "a" is a very small area at a distance "y" from the NA, then magnitude of the stress on area "a" equals (fb/c)(y). Remembering that stress is equal to a force distributed over an area, the force on the area "a" is equal to the stress times that area, or (fb/c)(y)(a). And, the moment of the force on area "a" about the NA is equal to (fb/c)(y)(a)(y).

Now, if all of the moments of all of the forces on all of the tiny "a" areas across the face of the beam are added together we arrive at an expression of the internal moment resistance of a beam;

$$M = \text{Sum } (f/c)ay^2 \text{ or } M = (f/c) \text{ Sum } ay^2$$

In this case, the value of (f/c) is a constant for any beam and Sum ay^2 is the second moment of the area, better known as the Moment of Inertia. Therefore, substituting items that we know, the equation becomes:

$$M = (fb/c)\ I$$

or rearranging

$$M = (I/c)\ fb$$

or

$$M = S*fb$$

(I/c) is known as the Section Modulus (S) and is also tabulated.

Bending Stresses Review

The flexure formula:

$$M = (f/c)\ I \text{ or } M = (I/c)\ f$$

M is the internal moment in a beam created by the loads and reactions. It is numerically equal to the external moment at any section. Its value can be taken directly from the moments diagram. M is in inch-pounds (or inch-kips).

f is the stress at any distance "c" from the NA. If "c" is taken as the distance from the NA to the most remote fibre, then the stress "f" is the greatest stress. c is in inches; f is psi (or ksi).

I is the Moment of inertia of the beam. I is in $inches^4$.

S is the Section Modulus. It is (I/c.) and has the units of $inches^3$.

The flexure formula is useful in the following ways:

- M = (Fb/c) I to find the resisting moment of a beam
- fb = M c /I to find the stress in a beam
- I / c = Fb to select a beam
- Sreq = M / Fb to select a beam

The last formula is the most useful. The value of I/c, or S, is constant for any given beam and is utilised so often that its value is listed in any table of beam properties.

Shearing Stresses

For small members in shear, it is customary to assume shear is uniformly distributed over the entire area. However, this assumption is not permissible for beam cross sections. The vertical shearing stress at any point in a beam may be determined from the horizontal shearing stress at that cross-section.

- The infinitesimally small particle “A”, whose dimensions are a, b, and 1, lies in a plane subject to shear.
- fs represents the vertical shearing stress and fv represents the horizontal shearing stress.
- fsb1 is the total shearing force on one vertical surface (force is equal to stress fs times the area b1); since the particle is in equilibrium an equal and opposite force must be acting on the other parallel face.
- Similarly, fva1 represents the total shearing force on each of the horizontal planes.
- The particle is in equilibrium, therefore the sum of the moments about any axis must equal zero. Take moments about axis “O”:

This shows that at any point, the horizontal shearing stress is equal to the vertical shearing stress.

The general shear formula will find the shearing stress at any point in a beam. It is:

$$fv = V A' y / I b$$

V = the absolute value of vertical shear from the shear diagram at the point where shear is being investigated.

A’ = the area above or below the horizontal plane being investigated.

y = the distance from the NA to the centroid of A’.

I = the moment of inertia of the entire section

b = the width of beam at the point where shear is being investigated.

Before deriving the formula, we will look at an application. If we wanted to find the maximum stress in the glue-line for this 2 x 10 composite beam:

- First draw the V diagram.
- Next, determine the geometry and associated properties of the beam and the appropriate value for y.
- Then plug into the equation to determine the stress at the glue-line.
- In this case, the stress would be 3000# (4")(2")(3") / (1/12)(2")(10^3)(2") or 216 psi.

Note: This is not a *bending* stress, it is a shearing stress.

Chapter 10

Earthquake Engineering

Earthquake engineering is the scientific field concerned with protecting society, the natural and the man-made environment from earthquakes by limiting the seismic risk to socioeconomically acceptable levels. Traditionally, it has been narrowly defined as the study of the behaviour of structures and geo-structures subject to seismic loading, thus considered as a subset of both structural and geotechnical engineering. However, the tremendous costs experienced in recent earthquakes have led to an expansion of its scope to encompass disciplines from the wider field of civil engineering and from the social sciences, especially sociology, political sciences, economics and finance. The main objectives of earthquake engineering are:

- Foresee the potential consequences of strong earthquakes on urban areas and civil infrastructure.
- Design, construct and maintain structures to perform at earthquake exposure up to the expectations and in compliance with building codes.

A properly engineered structure does not necessarily have to be extremely strong or expensive. It has to be properly designed to withstand the seismic effects while sustaining an acceptable level of damage.

Seismic Loading

Seismic loading means application of an earthquake-generated excitation to a structure (or geo-structure). It happens at contact surfaces of a structure either with the ground, or with adjacent structures, or with gravity waves from tsunami.

Seismic Performance

Earthquake or seismic performance defines a structure's ability to sustain its due functions, such as its safety and serviceability, *at* and *after* a particular earthquake exposure. A structure is, normally, considered *safe* if it does not endanger the lives and well-being of those in or around it by partially or completely collapsing. A structure may be considered *serviceable* if it is able to fulfill its operational functions for which it was designed.

Basic concepts of the earthquake engineering, implemented in the major building codes, assume that a building should survive a rare, very severe earthquake by sustaining significant damage but without globally collapsing. On the other hand, it should remain operational for more frequent, but less severe seismic events.

Seismic Performance Assessment

Engineers need to know the quantified level of the actual or anticipated seismic performance associated with the direct damage to an individual building subject to a specified ground shaking. Such an assessment may be performed either experimentally or analytically.

Experimental Assessment

Experimental evaluations are expensive tests that are typically done by placing a (scaled) model of the structure on a shake-table that simulates the earth shaking and observing its behaviour. Such kinds of experiments were first performed more than a century ago. Still only recently has it become possible to perform 1:1 scale testing on full structures. Due to the costly nature of such tests, they tend to be used mainly for understanding the seismic behaviour of structures, validating modells and verifying analysis methods. Thus, once properly validated, computational modells and numerical procedures tend to carry the major burden for the seismic performance assessment of structures.

Analytical/Numerical Assessment

Seismic performance assessment or, simply, seismic structural analysis is a powerful tool of earthquake engineering which utilises detailed modelling of the structure together with methods of structural analysis to gain a better understanding of seismic performance of building and non-building structures. The technique as a formal concept is a relatively recent development. In general, seismic structural analysis is based on the methods of structural dynamics. For decades, the most prominent instrument of seismic analysis has been the earthquake response spectrum method which, also, contributed to the proposed building code's concept of today.

However, such methods are good only for linear elastic systems, being largely unable to model the structural behaviour when damage (i.e., non-linearity) appears. Numerical *step-by-step integration* proved to be a more effective method of analysis for multi-degree-of-freedom structural systems with significant non-linearity under a transient process of ground motion excitation.

Basically, numerical analysis is conducted in order to evaluate the seismic performance of buildings. Performance evaluations are generally carried out by using nonlinear static pushover analysis or nonlinear time-history analysis.

In such analyses, it is essential to achieve accurate nonlinear modelling of structural components such as beams, columns, beam-column joints, shear walls etc. Thus, experimental results play an important role in determining the modelling parametres of individual components, especially those that are subject to significant nonlinear deformations. The individual components are then assembled to create a full nonlinear model of the structure. Thus created modells are analysed to evaluate the performance of buildings.

The capabilities of the structural analysis software are a major consideration in the above process as they restrict the possible component modells, the analysis methods available and, most importantly, the numerical robustness.

The latter becomes a major consideration for structures that venture into the nonlinear range and approach global or local collapse as the numerical solution becomes increasingly unstable and thus difficult to reach. There are several commercially available Finite Element Analysis software's such as CSI-SAP2000 and CSI-PERFORM-3D which can be used for the seismic performance evaluation of buildings. Moreover, there is research-based finite element analysis platforms such as Open Sees, RUAUMOKO and the older DRAIN-2D/3D, several of which are now open source.

Research for Earthquake Engineering

Research for earthquake engineering means both field and analytical investigation or experimentation intended for discovery and scientific explanation of earthquake engineering related facts, revision of conventional concepts in the light of new findings, and practical application of the developed theories. The National Science Foundation (NSF) is the main United States government agency that supports fundamental research and education in all fields of earthquake engineering. In particular, it focuses on experimental, analytical, and computational research on design and performance enhancement of structural systems.

Figure: *Shake-table testing of Friction Pendulum Bearings at EERC*

The Earthquake Engineering Research Institute (EERI) is a leader in dissemination of earthquake engineering research related information both in the U.S. and globally. A definitive list of earthquake engineering research related shaking tables around the world may be found in Experimental Facilities for Earthquake Engineering Simulation Worldwide. The most prominent of them is now E-Defence Shake Table in Japan. Earthquake engineering research activities worldwide are mostly associated with the following centres:

- Earthquake Engineering Research Institute (EERI)
- Earthquake Engineering Research Centre
- Pacific Earthquake Engineering Research Centre (PEER)
- John A. Blume Earthquake Engineering Centre
- Consortium of Universities for Research in Earthquake Engineering (CUREE)
- Multidisciplinary Centre for Earthquake Engineering Research (MCEER)

- George E. Brown, Jr. Network for Earthquake Engineering Simulation (NEES)
- USGS Earthquake Hazards Program
- Office of Earthquake Engineering at Caltrans
- Earthquake Engineering Research Centre of Iceland
- Earthquake Engineering New Zealand
- Canadian Research Centres and Research Groups on Earthquake Engineering
- Hyogo Earthquake Engineering Research Centre
- Laboratory for Earthquake Engineering of NTUA
- Earthquakes and Earthquake Engineering in The Library of Congress
- International Institute of Earthquake Engineering and Seismology
- National Centre for Research on Earthquake Engineering

Major U.S. Research Programs

The NSF Hazard Mitigation and Structural Engineering program (HMSE) supports research on new technologies for improving the behaviour and response of structural systems subject to earthquake hazards; fundamental research on safety and reliability of constructed systems; innovative developments in analysis and model based simulation of structural behaviour and response including soil-structure interaction; design concepts that improve structure performance and flexibility; and application of new control techniques for structural systems.

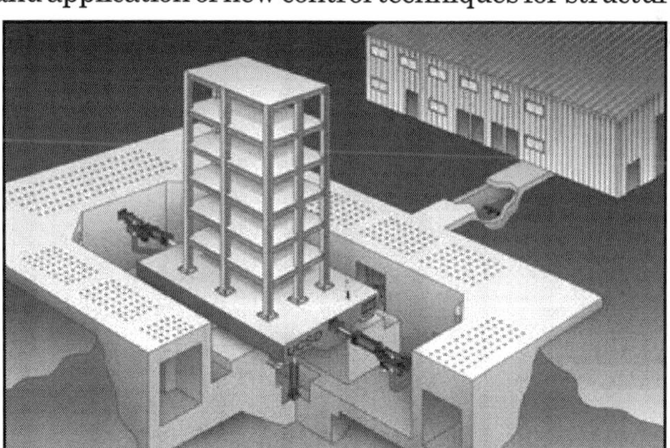

Figure : *Large High Performance Outdoor Shake Table, UCSD, NEES network.*

NSF also supports the George E. Brown, Jr. Network for Earthquake Engineering Simulation (NEES) that advances knowledge discovery and innovation for earthquakes and tsunami loss reduction of the nation's civil infrastructure, and new experimental simulation techniques and instrumentation.

The NEES network features 14 geographically-distributed, shared-use laboratories that support several types of experimental work: geotechnical centrifuge research, shake-table tests, large-scale structural testing, tsunami wave basin experiments, and field site research. Participating universities include: Cornell University; Lehigh University; Oregon State University; Rensselaer Polytechnic Institute; University at Buffalo, SUNY; University of California, Berkeley; University of California, Davis; University of California, Los Angeles; University of California, San Diego; University of California, Santa Barbara; University of Illinois, Urbana-Champaign; University of Minnesota; University of Nevada, Reno; and the University of Texas, Austin.

The equipment sites (labs) and a central data repository are connected to the global earthquake engineering community via the NEES hub website. The NEES website is powered by HUBzero software developed at Purdue University for nanoHUB specifically to help the scientific community share resources and collaborate. The cyber infrastructure, connected via Internet 2, provides interactive simulation tools, a simulation tool development area, a curated central data repository, animated presentations, user support, telepresence, mechanism for uploading and sharing resources and statistics about users, and usage patterns.

This cyber infrastructure allows researchers to: securely store, organize and share data within a standardised framework in a central location; remotely observe and participate in experiments through the use of synchronised real-time data and video; collaborate with colleagues to facilitate the planning, performance, analysis, and publication of research experiments; and conduct computational and hybrid simulations that may combine the results of multiple distributed experiments and link physical experiments with computer simulations to enable the investigation of overall system performance.

These resources jointly provide the means for collaboration and discovery to improve the seismic design and performance of civil and mechanical infrastructure systems.

Earthquake Simulation

The very first earthquake simulations were performed by statically applying some *horizontal inertia forces* based on scaled peak ground accelerations to a mathematical model of a building. With the further development of computational technologies, static approaches began to give way to dynamic ones.

Dynamic experiments on building and non-building structures may be physical, like shake-table testing, or virtual ones. In both cases, to verify a structure's expected seismic performance, some researchers prefer to deal with so called "real time-histories" though the last cannot be "real" for a hypothetical earthquake specified by either a building code or by some particular research requirements. Therefore, there is a strong incentive to engage an earthquake simulation which is the seismic input that possesses only essential features of a real event. Sometimes, earthquake simulation is understood as a recreation of local effects of a strong earth shaking.

Structure Simulation

Theoretical or experimental evaluation of anticipated seismic performance mostly requires a structure simulation which is based on the concept of structural likeness or similarity. Similarity is some degree of analoguy or resemblance between two or more objects. The notion of similarity rests either on exact or approximate repetitions of patterns in the compared items.

Figure: *Concurrent experiments with two building models which are kinematically equivalent to a real prototype.*

In general, a building model is said to have similarity with the real object if the two share *geometric similarity*, *kinematic similarity* and *dynamic similarity*. The most vivid and effective type of similarity is the *kinematic* one. *Kinematic similarity* exists when the paths and velocities of moving particles of a model and its prototype are similar. The ultimate level of *kinematic similarity* is *kinematic equivalence* when, in the case of earthquake engineering, time-histories of each story lateral displacements of the model and its prototype would be the same.

Seismic Vibration Control

Seismic vibration control is a set of technical means aimed to mitigate seismic impacts in building and non-building structures. All seismic vibration control devices may be classified as *passive*, *active* or *hybrid* where:

- *passive control devices* have no feedback capability between them, structural elements and the ground;
- *active control devices* incorporate real-time recording instrumentation on the ground integrated with earthquake input processing equipment and actuators within the structure;
- *hybrid control devices* have combined features of active and passive control systems.

When ground seismic waves reach up and start to penetrate a base of a building, their energy flow density, due to reflections, reduces dramatically: usually, up to 90%. However, the remaining portions of the incident waves during a major earthquake still bear a huge devastating potential.

After the seismic waves enter a superstructure, there are a number of ways to control them in order to soothe their damaging effect and improve the building's seismic performance, for instance:

- to dissipate the wave energy inside a superstructure with properly engineered dampers;
- to disperse the wave energy between a wider range of frequencies;
- to absorb the resonant portions of the whole wave frequencies band with the help of so called *mass dampers*.

Devices of the last kind, abbreviated correspondingly as TMD for the tuned (*passive*), as AMD for the *active*, and as HMD for the *hybrid mass dampers*, have been studied and installed in high-rise buildings, predominantly in Japan, for a quarter of a century.

However, there is quite another approach: partial suppression of the seismic energy flow into the superstructure known as seismic or base isolation. For this, some pads are inserted into or under all major load-carrying elements in the base of the building which should substantially decouple a superstructure from its substructure resting on a shaking ground.

The first evidence of earthquake protection by using the principle of base isolation was discovered in Pasargadae, a city in ancient Persia, now Iran: it goes back to 6th century BCE. Below, there are some samples of seismic vibration control technologies of today.

Dry-stone Walls Control

People of Inca civilization were masters of the polished '*dry-stone walls*', called ashlar, where blocks of stone were cut to fit together tightly without any mortar. The Incas were among the best stonemasons the world has ever seen, and many junctions in their masonry were so perfect that even blades of grass could not fit between the stones. Peru is a highly seismic land, and for centuries the mortar-free construction proved to be apparently more earthquake-resistant than using mortar. The stones of the dry-stone walls built by the Incas could move slightly and resettle without the walls collapsing, a passive structural control technique employing both the principle of energy dissipation and that of suppressing resonant amplifications.

Lead Rubber Bearing

Lead Rubber Bearing or LRB is a type of base isolation employing a heavy damping. It was invented by Bill Robinson, a New Zealander.

Heavy damping mechanism incorporated in vibration control technologies and, particularly, in base isolation devices, is often considered a valuable source of suppressing vibrations thus enhancing a building's seismic performance. However, for the rather pliant systems such as base isolated structures, with a relatively low bearing stiffness but with a high damping, the so-called "damping force" may turn out the main pushing force at a strong earthquake. The video shows a Lead Rubber Bearing being tested at the UCSD Caltrans-SRMD facility. The bearing is made of rubber with a lead core. It was a uni-axial test in which the bearing was also under a full structure load. Many buildings and bridges, both in New Zealand and elsewhere, are protected with lead dampers and lead and rubber bearings. Te Papa Tongarewa, the national museum of New Zealand, and the New Zealand Parliament Buildings have been fitted with the bearings. Both are in Wellington, which sits on an active earthquake fault.

Tuned Mass Damper

Typically, the tuned mass dampers are huge concrete blocks mounted in skyscrapers or other structures and moved in opposition to the resonance frequency oscillations of the structures by means of some sort of spring mechanism.

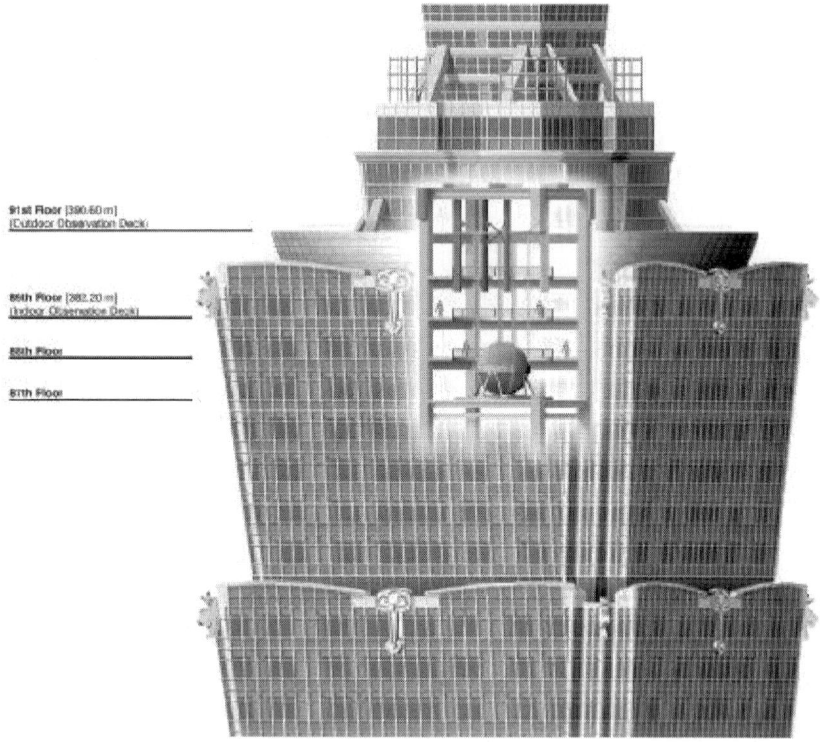

Figure: *Tuned mass damper in Taipei 101, the world's third tallest skyscraper*

Taipei 101 skyscraper needs to withstand typhoon winds and earthquake tremors common in its area of the Asia-Pacific.

For this purpose, a steel pendulum weighing 660 metric tons that serves as a tuned mass damper was designed and installed atop the structure.

Suspended from the 92nd to the 88th floor, the pendulums sways to decrease resonant amplifications of lateral displacements in the building caused by earthquakes and strong gusts.

Friction Pendulum Bearing

Friction Pendulum Bearing (FPB) is another name of Friction Pendulum System (FPS). It is based on three pillars:

- articulated friction slider;
- spherical concave sliding surface;
- enclosing cylinder for lateral displacement restraint.

Snapshot with the link to video clip of a shake-table testing of FPB system supporting a rigid building model is presented at the right.

Building Elevation Control

Building elevation control is a valuable source of vibration control of seismic loading. Pyramid-shaped skyscrapers continue to attract attention of architects and engineers because such structures promise a better stability against earthquakes and winds.

The elevation configuration can prevent buildings' resonant amplifications because a properly configured building disperses the shear wave energy between a wide range of frequencies.

Earthquake or wind quieting ability of the elevation configuration is provided by a specific pattern of multiple reflections and transmissions of vertically propagating shear waves, which are generated by breakdowns into homogeneity of story layers, and a taper.

Any abrupt changes of the propagating waves velocity result in a considerable dispersion of the wave energy between a wide ranges of frequencies thus preventing the resonant displacement amplifications in the building.

A tapered profile of a building is not a compulsory feature of this method of structural control. A similar resonance preventing effect can be also obtained by a proper *tapering* of other characteristics of a building structure, namely, its mass and stiffness.

As a result, the building elevation configuration techniques permit an architectural design that may be both attractive and functional.

Simple Roller Bearing

Simple roller bearing is a base isolation device which is intended for protection of various building and non-building structures against potentially damaging lateral impacts of strong earthquakes.

This metallic bearing support may be adapted, with certain precautions, as a seismic isolator to skyscrapers and buildings on soft ground. Recently, it has been employed under the name of *Metallic Roller Bearing* for a housing complex (17 stories) in Tokyo, Japan.

Springs-with-damper Base Isolator

Springs-with-damper base isolator installed under a three-story town-house, Santa Monica, California is shown on the photo taken prior to the 1994 Northridge earthquake exposure. It is a base isolation device conceptually similar to *Lead Rubber Bearing*.

***Figure** : Springs-with-damper close-up.*

One of two three-story town-houses like this, which was well instrumented for recording of both vertical and horizontal accelerations on its floors and the ground, has survived a severe shaking during the Northridge earthquake and left valuable recorded information for further study.

Hysteretic Damper

Hysteretic damper is intended to provide better and more reliable seismic performance than that of a conventional structure at the expense of the seismic input energy dissipation. There are four major groups of hysteretic dampers used for the purpose, namely:

- Fluid viscous dampers (FVDs)
- Metallic yielding dampers (MYDs)
- Viscoelastic dampers (VEDs)
- Friction dampers (FDs).

Each group of dampers has specific characteristics, advantages and disadvantages for structural applications.

Seismic Design

Seismic design is based on authorised engineering procedures, principles and criteria meant to design or retrofit structures subject to earthquake exposure. Those criteria are only consistent with the contemporary state of the knowledge about earthquake engineering structures. Therefore, a building design which exactly follows seismic code regulations does not guarantee safety against collapse or serious damage.

The price of poor seismic design may be enormous. Nevertheless, seismic design has always been a trial and error process whether it was based on physical laws or on empirical knowledge of the structural performance of different shapes and materials.

To practice seismic design, seismic analysis or seismic evaluation of new and existing civil engineering projects, an engineer should, normally, pass examination on *Seismic Principles* which, in the State of California, include:

- Seismic Data and Seismic Design Criteria
- Seismic Characteristics of Engineered Systems
- Seismic Forces
- Seismic Analysis Procedures
- Seismic Detailing and Construction Quality Control.

To build up complex structural systems, seismic design largely uses the same relatively small number of basic structural elements (to say nothing of vibration control devices) as any non-seismic design project.

Normally, according to building codes, structures are designed to "withstand" the largest earthquake of a certain probability that is likely to occur at their location. This means the loss of life should be minimised by preventing collapse of the buildings.

Seismic design is carried out by understanding the possible failure modes of a structure and providing the structure with appropriate strength, stiffness, ductility, and configuration to ensure those modes cannot occur.

Seismic Design Requirements

Seismic design requirements depend on the type of the structure, locality of the project and its authorities which stipulate applicable seismic design codes and criteria. For instance, California Department of Transportation's requirements called *The Seismic Design Criteria* (SDC) and aimed at the design of new bridges in California incorporate an innovative seismic performance based approach.

***Figure** : Metsamor Nuclear Power Plant in Armenia was closed after the 1988 destructive earthquake.*

The most significant feature in the SDC design philosophy is a shift from a *force-based assessment* of seismic demand to a *displacement-based assessment* of demand and capacity. Thus, the newly adopted displacement approach is based on comparing the *elastic displacement* demand to the *inelastic displacement* capacity of the primary structural components while ensuring a minimum level of inelastic capacity at all potential plastic hinge locations.

In addition to the designed structure itself, seismic design requirements may include a *ground stabilisation* underneath the structure: sometimes, heavily shaken ground breaks up which leads to collapse of the structure sitting upon it. The following topics should be of primary concerns: liquefaction; dynamic lateral earth pressures on retaining walls; seismic slope stability; earthquake-induced settlement.

Nuclear facilities should not jeopardise their safety in case of earthquakes or other hostile external events. Therefore, their seismic design is based on criteria far more stringent than those applying to non-nuclear facilities. The Fukushima I nuclear accidents and damage to other nuclear facilities that followed the 2011 Tôhoku earthquake and tsunami have, however, drawn attention to ongoing concerns over

Japanese nuclear seismic design standards and caused other many governments to re-evaluate their nuclear programs. Doubt has also been expressed over the seismic evaluation and design of certain other plants, including the Fessenheim Nuclear Power Plant in France.

Failure Modes

Failure mode is the manner by which an earthquake induced failure is observed. It, generally, describes the way the failure occurs. Though costly and time consuming, learning from each real earthquake failure remains a routine recipe for advancement in *seismic design* methods. Below, some typical modes of earthquake-generated failures are presented.

The lack of reinforcement coupled with poor mortar and inadequate roof-to-wall ties can result in substantial damage to a unreinforced masonry building. Severely cracked or leaning walls are some of the most common earthquake damage. Also hazardous is the damage that may occur between the walls and roof or floor diaphragms. Separation between the framing and the walls can jeopardize the vertical support of roof and floor systems.

Soft story effect. Absence of adequate shear walls on the ground level caused damage to this structure. A close examination of the image reveals that the rough board siding, once covered by a brick veneer, has been completely dismantled from the studwall. Only the rigidity of the floor above combined with the support on the two hidden sides by continuous walls, not penetrated with large doors as on the street sides, is preventing full collapse of the structure.

Soil liquefaction. In the cases where the soil consists of loose granular deposited materials with the tendency to develop excessive hydrostatic pore water pressure of sufficient magnitude and compact, liquefaction of those loose saturated deposits may result in non-uniform settlements and tilting of structures. This caused major damage to thousands of buildings in Niigata, Japan during the 1964 earthquake.

Landslide rock fall. A landslide is a geological phenomenon which includes a wide range of ground movement, including rock falls. Typically, the action of gravity is the primary driving force for a landslide to occur though in this case there was another contributing factor which affected the original slope stability: the landslide required an *earthquake trigger* before being released.

Pounding against adjacent building. This is a photograph of the collapsed five-story tower, St. Joseph's Seminary, Los Altos, California which resulted in one fatality. During Loma Prieta earthquake, the tower pounded against the independently vibrating adjacent building behind. A possibility of pounding depends on both buildings' lateral displacements which should be accurately estimated and accounted for.

At Northridge earthquake, the Kaiser Permanent concrete frame office building had joints completely shattered, revealing inadequate confinement steel, which resulted in the second story collapse. In the transverse direction, composite end shear walls, consisting of two wythes of brick and a layer of shotcrete that carried the lateral load, peeled apart because of inadequate through-ties and failed.

7 story reinforced concrete buildings on steep slope collapse due to the following :

- Improper construction site on a foothill.
- Poor detailing of the reinforcement (lack of concrete confinement in the columns and at the beam-column joints, inadequate splice length).
- Seismically weak soft story at the first floor.
- Long cantilevers with heavy dead load.

Sliding off foundations effect of a relatively rigid residential building structure during 1987 Whittier Narrows earthquake. The magnitude 5.9 earthquake pounded the Garvey West Apartment building in Monterey Park, California and shifted its superstructure about 10 inches to the east on its foundation.

If a superstructure is not mounted on a base isolation system, its shifting on the basement should be prevented.

Reinforced concrete column burst at Northridge earthquake due to insufficient shear reinforcement mode which allows main reinforcement to buckle outwards. The deck unseated at the hinge and failed in shear. As a result, the La Cienega-Venice underpass section of the 10 Freeway collapsed.

Loma Prieta earthquake: side view of reinforced concrete support-columns failure which trigged the upper deck collapse onto the lower deck of the two-level Cypress viaduct of Interstate Highway 880, Oakland, CA.

Retaining wall failure at Loma Prieta earthquake in Santa Cruz Mountains area: prominent northwest trending extensional cracks up to 12 cm (4.7 in) wide in the concrete spillway to Austrian Dam, the north abutment.

Ground shaking triggered soil liquefaction in a subsurface layer of sand, producing differential lateral and vertical movement in an overlying carapace of unliquified sand and silt. This mode of ground failure, termed lateral spreading, is a principal cause of liquefaction-related earthquake damage.

Severely damaged building of Agriculture Development Bank of China after 2008 Sichuan earthquake: most of the beams and pier columns are sheared. Large diagonal cracks in masonry and veneer are due to in-plane loads while abrupt settlement of the right end of the building should be attributed to a landfill which may be hazardous even without any earthquake, see video footage at.

Twofold tsunami impact: sea waves hydraulic pressure and inundation. Thus, 2004 Indian Ocean earthquake of December 26, 2004, with the epicentre off the west coast of Sumatra, Indonesia, triggered a series of devastating tsunamis, killing more than 225,000 people in eleven countries by inundating surrounding coastal communities with huge waves up to 30 metres (100 feet) high.

Earthquake-Resistant construction

Earthquake construction means implementation of seismic design to enable building and non-building structures to live through the anticipated earthquake exposure up to the expectations and in compliance with the applicable building codes.

Design and construction are intimately related. To achieve a good workmanship, detailing of the members and their connections should be, possibly, simple. As any construction in general, earthquake construction is a process that consists of the building, retrofitting or assembling of infrastructure given the construction materials available.

The destabilising action of an earthquake on constructions may be *direct* (seismic motion of the ground) or *indirect* (earthquake-induced landslides, soil liquefaction and waves of tsunami).

A structure might have all the appearances of stability, yet offer nothing but danger when an earthquake occurs. The crucial fact is that, for safety, earthquake-resistant construction techniques are as important as quality control and using correct materials. *Earthquake contractor* should be registered in the state of the project location, bonded and insured.

To minimize possible losses, construction process should be organised with keeping in mind that earthquake may strike any time prior to the end of construction.

Each construction project requires a qualified team of professionals who understand the basic features of seismic performance of different structures as well as construction management.

Adobe Structures

Around thirty percent of the world's population lives or works in earth-made construction. Adobe type of mud bricks is one of the oldest and most widely used building materials. The use of adobe is very common in some of the world's most hazard-prone regions, traditionally across Latin America, Africa, Indian subcontinent and other parts of Asia, Middle East and Southern Europe. Adobe buildings are considered very vulnerable at strong quakes. However, multiple ways of seismic strengthening of new and existing adobe buildings are available, see, e.g.

Key factors for the improved seismic performance of adobe construction are:

- Quality of construction.
- Compact, box-type layout.
- Seismic reinforcement.

Limestone and Sandstone Structures

Limestone is very common in architecture, especially in North America and Europe. Many landmarks across the world, including the pyramids in Egypt, are made of limestone. Many medieval churches and castles in Europe are made of limestone and sandstone masonry. They are the long-lasting materials but their rather heavy weight is not beneficial for adequate seismic performance. Application of modern technology to seismic retrofitting can enhance the survivability of unreinforced masonry structures. As an example, from 1973 to 1989, the Salt Lake City and County Building in Utah was exhaustively renovated and repaired with an emphasis on preserving historical accuracy in appearance. This was done in concert with a seismic upgrade that placed the weak sandstone structure on base isolation foundation to better protect it from earthquake damage.

Timber Frame Structures

Timber framing dates back thousands of years, and has been used in many parts of the world during various periods such as ancient Japan, Europe and medieval England in localities where timber was in good supply and building stone and the skills to work it were not.

The use of timber framing in buildings provides their complete skeletal framing which offers some structural benefits as the timber frame, if properly engineered, lends itself to better *seismic survivability*.

Light-frame Structures

***Figure** : A two-story wooden-frame for a residential building structure.*

Light-frame structures usually gain seismic resistance from rigid plywood shear walls and wood structural panel diaphragms. Special provisions for seismic load-resisting systems for all engineered wood structures requires consideration of diaphragm ratios, horizontal and vertical diaphragm shears, and connector/fastener values. In addition, collectors, or drag struts, to distribute shear along a diaphragm length are required.

Reinforced Masonry Structures

A construction system where steel reinforcement is embedded in the mortar joints of masonry or placed in holes and after filled with concrete or grout is called reinforced masonry.

***Figure** : Reinforced hollow masonry wall.*

The devastating 1933 Long Beach earthquake revealed that masonry construction should be improved immediately. Then, the California State Code made the reinforced masonry mandatory.

There are various practices and techniques to achieve reinforced masonry. The most common type is the reinforced hollow unit masonry. The effectiveness of both vertical and horizontal reinforcement strongly depends on the type and quality of the masonry, i.e. masonry units and mortar.

To achieve a ductile behaviour of masonry, it is necessary that the shear strength of the wall is greater than the tensile strength of reinforcement to ensure a kind of bending failure.

Reinforced Concrete Structures

Reinforced concrete is concrete in which steel reinforcement bars (rubbers) or fibres have been incorporated to strengthen a material that would otherwise be brittle. It can be used to produce beams, columns, floors or bridges. Prestressed concrete is a kind of reinforced concrete used for overcoming concrete's natural weakness in tension. It can be applied to beams, floors or bridges with a longer span than is practical with ordinary reinforced concrete. Prestressing tendons (generally of high tensile steel cable or rods) are used to provide a clamping load which produces a compressive stress that offsets the tensile stress that the concrete compression member would, otherwise, experience due to a bending load.

To prevent catastrophic collapse in response earth shaking (in the interest of life safety), a traditional reinforced concrete frame should have ductile joints. Depending upon the methods used and the imposed seismic forces, such buildings may be immediately usable, require extensive repair, or may have to be demolished.

Prestressed Structures

Prestressed structure is the one whose overall integrity, stability and security depend, primarily, on a *prestressing*. *Prestressing* means the intentional creation of permanent stresses in a structure for the purpose of improving its performance under various service conditions.

There are the following basic types of prestressing:

- Pre-compression (mostly, with the own weight of a structure)
- Pretensioning with high-strength embedded tendons
- Posttensioning with high-strength bonded or unbounded tendons.

Figure : Prestressed concrete cable-stayed bridge over Yangtze river.

Today, the concept of prestressed structure is widely engaged in design of buildings, underground structures, TV towers, power stations, floating storage and offshore facilities, nuclear reactor vessels, and numerous kinds of bridge systems. A beneficial idea of *prestressing* was, apparently, familiar to the ancient Rome architects; look, e.g., at the tall attic wall of Colosseum working as a stabilising device for the wall piers beneath.

Steel Structures

Steel structures are considered mostly earthquake resistant but their resistance should never be taken for granted. A great number of welded steel moment frame buildings, which looked earthquake-proof, surprisingly experienced brittle behaviour and were hazardously damaged in the 1994 Northridge earthquake. After that, the Federal Emergency Management Agency (FEMA) initiated development of repair techniques and new design approaches to minimize damage to steel moment frame buildings in future earthquakes.

For structural steel seismic design based on Load and Resistance Factor Design (LRFD) approach, it is very important to assess ability of a structure to develop and maintain its bearing resistance in the inelastic range. A measure of this ability is ductility, which may be observed in a *material itself*, in a *structural element*, or to a *whole structure*.

As a consequence of Northridge earthquake experience, all pre-qualified connection details and design methods contained in the building codes of that time have been rescinded. The new provisions stipulated that new designs be substantiated by testing or by use of test-verified calculations.

Prediction of Earthquake Losses

Earthquake loss estimation is usually defined as a *Damage Ratio* (DR) which is a ratio of the earthquake damage repair cost to the total value of a building. *Probable Maximum Loss* (PML) is a common term used for earthquake loss estimation, but it lacks a precise definition. In 1999, ASTM E2026 'Standard Guide for the Estimation of Building Damageability in Earthquakes' was produced in order to standardize the nomenclature for seismic loss estimation, as well as establish guidelines as to the review process and qualifications of the reviewer.

Earthquake loss estimations are also referred to as *Seismic Risk Assessments*. The risk assessment process generally involves determining the probability of various ground motions coupled with the vulnerability or damage of the building under those ground motions. The results are defined as a percent of building replacement value.

Environmental Engineering

Environmental engineering is the application of science and engineering principles to improve the natural environment (air, water, and/or land resources), to provide healthy water, air, and land for human habitation (house or home) and for other organisms, and to remediate polluted sites. It involves waste water management and air pollution control, recycling, waste disposal, radiation protection, industrial hygiene, environmental sustainability, and public health issues as well as a knowledge of environmental engineering law. It also includes studies on the environmental impact of proposed construction projects.

Environmental engineers conduct hazardous-waste management studies to evaluate the significance of such hazards, advise on treatment and containment, and develop regulations to prevent mishaps. Environmental engineers also design municipal water supply and industrial wastewater treatment systems as well as address local and worldwide environmental issues such as the effects of acid rain, global warming, ozone depletion, water pollution and air pollution from automobile exhausts and industrial sources. At many universities, Environmental Engineering programs follow either the Department of Civil Engineering or The Department of Chemical Engineering at Engineering faculties. Environmental "civil" engineers focus on hydrology, water resources management, bioremediation, and water treatment plant design. Environmental "chemical" engineers, on the other hand, focus on environmental chemistry, advanced air and water treatment technologies and separation processes.

Additionally, engineers are more frequently obtaining specialised training in law (J.D.) and are utilising their technical expertise in the practices of Environmental engineering law. About four percent of environmental engineers go on to obtain Board Certification in their speciality area(s) of environmental engineering (Board Certified Environmental Engineer or BCEE). Most jurisdictions also impose licensing and registration requirements.

Development of Environmental Engineering

Ever since people first recognised that their health and well-being were related to the quality of their environment, they have applied thoughtful principles to attempt to improve the quality of their environment. The ancient Harappan civilization utilised early sewers in some cities. The Romans constructed aqueducts to prevent drought and to create a clean, healthful water supply for the metropolis of Rome. In the 15th century, Bavaria created laws restricting the development and degradation of alpine country that constituted the region's water supply.

The field emerged as a separate environmental discipline during the middle third of the 20th century in response to widespread public concern about water and pollution and increasingly extensive environmental quality degradation. However, its roots extend back to early efforts in public health engineering. Modern environmental engineering began in London in the mid 19th century when Joseph Bazalgette designed the first major sewerage system that reduced the incidence of waterborne diseases such as cholera. The introduction of drinking water treatment and sewage treatment in industrialised countries reduced waterborne diseases from leading causes of death to rarities.

In many cases, as societies grew, actions that were intended to achieve benefits for those societies had longer-term impacts which reduced other environmental qualities. One example is the widespread application of DDT to control agricultural pests in the years following World War II. While the agricultural benefits were outstanding and crop yields increased dramatically, thus reducing world hunger substantially, and malaria was controlled better than it ever had been, numerous species were brought to the verge of extinction due to the impact of the DDT on their reproductive cycles. The story of DDT as vividly told in Rachel Carson's "Silent Spring" is considered to be the birth of the modern environmental movement and the development of the modern field of "environmental engineering." Conservation movements and laws restricting public actions that would harm the environment have been developed by various societies for millennia. Notable examples are the laws decreeing the construction of sewers in London and Paris in the 19th century and the creation of the U.S. national park system in the early 20th century.

Scope of Environmental Engineering

Briefly speaking, the main task of environmental engineers is to protect public health by protecting (from further degradation), preserving (the present condition of), and enhancing the environment. Environmental engineering is the application of science and engineering principles to the environment. Some consider environmental engineering to include the development of sustainable processes. There are several divisions of the field of environmental engineering.

Environmental Impact Assessment and Mitigation

In this division, engineers and scientists use a systemic identification and evaluation process to assess the potential impacts of a proposed project, plans, programs, policies, or legislative actions upon the physical-chemical, biological, cultural, and socioeconomic components on environmental conditions. They apply scientific and engineering principles to evaluate if there are likely to be any adverse impacts to water quality, air quality, habitat quality, flora and fauna, agricultural capacity, traffic impacts, social impacts, ecological impacts, noise impacts, visual (landscape) impacts, etc. If impacts are expected, they then develop mitigation measures to limit or prevent such impacts. An example of a mitigation measure would be the creation of wetlands in a nearby location to mitigate the filling in of wetlands necessary for a road development if it is not possible to reroute the road.

The practice of environmental assessment was intitiated on January 1, 1970, the effective date of the National Environmental Policy Act (NEPA) in the United States. Since that time, more than 100 developing and developed nations either have planned specific analoguous laws or have adopted procedure used elsewhere. NEPA is applicable to all federal agencies in the United States.

Water Supply and Treatment

Engineers and scientists work to secure water supplies for potable and agricultural use. They evaluate the water balance within a watershed and determine the available water supply, the water needed for various needs in that watershed, the seasonal cycles of water movement through the watershed and they develop systems to store, treat, and convey water for various uses. Water is treated to achieve water quality objectives for the end uses. In the case of potable water supply, water is treated to minimize the risk of infectious disease transmission, the risk of noninfectious illness, and to create a palatable water flavour. Water distribution systems are designed and built to provide adequate water pressure and flow rates to meet various end-user needs such as domestic use, fire suppression, and irrigation.

Wastewater Conveyance and Treatment

Most urban and many rural areas no longer discharge human waste directly to the land through outhouse, septic, and/or honey bucket systems, but rather deposit such waste into water and convey it from households via sewer systems. Engineers and scientists develop collection and treatment systems to carry this waste material away from where people live and produce the waste and discharge it into the environment. In developed countries, substantial resources are applied to the treatment and detoxification of this waste before it is discharged into a river, lake, or ocean system. Developing nations are striving to obtain the resources to develop such systems so that they can improve water quality in their surface waters and reduce the risk of waterborne infectious disease.

There are numerous wastewater treatment technologies. A wastewater treatment train can consist of a primary clarifier system to remove solid and floating materials, a secondary treatment system consisting of an aeration basin followed by flocculation and sedimentation or an activated sludge system and a secondary clarifier, a tertiary biological nitrogen removal system, and a final disinfection process. The aeration basin/activated sludge system removes organic material by growing bacteria (activated sludge). The secondary clarifier removes the activated sludge from the water. The tertiary system, although not always included due to costs, is becoming more prevalent to remove nitrogen and phosphorus and to disinfect the water before discharge to a surface water stream or ocean outfall.

Air Quality Management

Engineers apply scientific and engineering principles to the design of manufacturing and combustion processes to reduce air pollutant emissions to acceptable levels. Scrubbers, electrostatic precipitators, catalytic converters, and various other processes are utilised to remove particulate matter, nitrogen oxides, sulphur oxides, volatile organic compounds (VOC), reactive organic gases (ROG) and other air pollutants from flue gases and other sources prior to allowing their emission to the atmosphere.

Scientists have developed air pollution dispersion modells to evaluate the concentration of a pollutant at a receptor or the impact on overall air quality from vehicle exhausts and industrial flue gas stack emissions. To some extent, this field overlaps the desire to decrease carbon dioxide and other greenhouse gas emissions from combustion processes.

Other Applications

- Environmental policy and regulation development
- Contaminated land management and site remediation
- Environment, Health and Safety
- Hazardous waste management
- Natural resource management
- Noise pollution
- Risk assessment
- Solid waste management.

Education

Courses aimed at developing graduates with specific skills in environmental systems or environmental technology are becoming more common and fall into three broads classes:

- *Environmental Engineering or Environmental Systems* courses oriented towards a civil engineering approach in which structures and the landscape are constructed to blend with or protect the environment;
- *Environmental chemistry, sustainable chemistry* or *environmental chemical engineering* courses oriented towards understanding the effects (good and bad) of chemicals in the environment. Such awards can focus on mining processes, pollutants and commonly also cover biochemical processes;
- *Environmental technology* courses oriented towards producing electronic or electrical graduates capable of developing devices and artefacts able to monitor, measure, model and control environmental impact, including monitoring and managing energy generation from renewable sources.

Geotechnical Engineering

Geotechnical engineering is the branch of civil engineering concerned with the engineering behaviour of earth materials. Geotechnical engineering is important in civil engineering, but is also used by military, mining, petroleum, or any other engineering concerned with construction on or in the ground. Geotechnical engineering uses principles of soil mechanics and rock mechanics to investigate subsurface conditions and materials; determine the relevant physical/mechanical and chemical properties of these materials; evaluate stability of natural slopes and man-made soil deposits; assess risks posed by site conditions; design earthworks and structure foundations; and monitor site conditions, earthwork and foundation construction.

A typical geotechnical engineering project begins with a review of project needs to define the required material properties. Then follows a site investigation of soil, rock, fault distribution and bedrock properties on and below an area of interest to determine their engineering properties including how they will interact with, on or in a proposed construction. Site investigations are needed to gain an understanding of the area in or on which the engineering will take place. Investigations can include the assessment of the risk to humans, property and the environment from natural hazards such as earthquakes, landslides, sinkholes, soil liquefaction, debris flows and rockfalls.

Ground Improvement refers to a technique that improves the engineering properties of the soil mass treated. Usually, the properties that are modified are shear strength, stiffness and permeability. Ground improvement has developed into a sophisticated tool to support foundations for a wide variety of structures. Properly applied, i.e. after giving due to consideration to the nature of the ground being improved and the type and sensitivity of the structures being built, ground improvement often reduces directs costs and saves time.

A geotechnical engineer then determines and designs the type of foundations, earthworks, and/or pavement subgrades required for the intended man-made structures to be built. Foundations are designed and constructed for structures of various sizes such as high-rise buildings, bridges, medium to large commercial buildings, and smaller structures where the soil conditions do not allow code-based design.

Foundations built for above-ground structures include shallow and deep foundations. Retaining structures include earth-filled dams and retaining walls. Earthworks include embankments, tunnels, dikes, levees, channels, reservoirs, deposition of hazardous waste and sanitary landfills.

Geotechnical engineering is also related to coastal and ocean engineering. Coastal engineering can involve the design and construction of wharves, marinas, and jetties. Ocean engineering can involve foundation and anchor systems for offshore structures such as oil platforms. The fields of geotechnical engineering and engineering geology are closely related, and have large areas of overlap. However, the field of geotechnical engineering is a speciality of engineering, where the field of engineering geology is a speciality of geology.

History

Humans have historically used soil as a material for flood control, irrigation purposes, burial sites, building foundations, and as construction material for buildings. First activities were linked to irrigation and flood control, as demonstrated by traces of dykes, dams, and canals dating back to at least 2000 BCE that were found in ancient Egypt, ancient Mesopotamia and the Fertile Crescent, as well as around the early settlements of Mohenjo Daro and Harappa in the Indus valley. As the cities expanded, structures were erected supported by formalised foundations; Ancient Greeks notably constructed pad footings and strip-and-raft foundations. Until the 18 th century, however, no theoretical basis for soil design had been developed and the discipline was more of an art than a science, relying on past experience.

Several foundation-related engineering problems, such as the Leaning Tower of Pisa, prompted scientists to begin taking a more scientific-based approach to examining the subsurface. The earliest advances occurred in the development of earth pressure theories for the construction of retaining walls. Henri Gautier, a French Royal Engineer, recognised the "natural slope" of different soils in 1717, an idea later known as the soil's angle of repose. A rudimentary soil classification system was also developed based on a material's unit weight, which is no longer considered a good indication of soil type.

The application of the principles of mechanics to soils was documented as early as 1773 when Charles Coulomb (a physicist, engineer, and army Captain) developed improved methods to determine the earth pressures against military ramparts. Coulomb observed that, at failure, a distinct slip plane would form behind a sliding retaining wall and he suggested that the maximum shear stress on the slip plane, for design purposes, was the sum of the soil cohesion, *c*, and friction $\sigma\tan(\phi)$, where is the normal stress on the slip plane and is the friction angle of the soil. By combining Coulomb's theory with Christian Otto Mohr's 2D stress state, the theory became known as Mohr-Coulomb theory. Although it is now recognised that precise determination of cohesion is impossible because *c* is not a fundamental soil property, the Mohr-Coulomb theory is still used in practice today.

In the 19th century Henry Darcy developed what is now known as Darcy's Law describing the flow of fluids in porous media. Joseph Boussinesq (a mathematician and physicist) developed theories of stress distribution in elastic solids that proved useful for estimating stresses at depth in the ground; William Rankine, an engineer and physicist, developed an alternative to Coulomb's earth pressure theory. Albert Atterberg developed the clay consistency indices that are still used today for soil classification. Osborne Reynolds recognised in 1885 that shearing causes volumetric dilation of dense and contraction of loose granular materials.

Modern geotechnical engineering is said to have begun in 1925 with the publication of *Erdbaumechanik* by Karl Terzaghi (a civil engineer and geologist). Considered by many to be the father of modern soil mechanics and geotechnical engineering, Terzaghi developed the principle of effective stress, and demonstrated that the shear strength of soil is controlled by effective stress. Terzaghi also developed the framework for theories of bearing capacity of foundations, and the theory for prediction of the rate of settlement of clay layers due to consolidation. In his 1948 book, Donald Taylor recognised that interlocking and dilation of densely packed particles contributed to the peak strength of a soil. The interrelationships between volume change behaviour (dilation, contraction, and consolidation) and shearing behaviour were all connected via the theory of plasticity using critical state soil mechanics by Roscoe, Schofield, and Wroth with the publication of "On the Yielding of Soils" in 1958. Critical state soil mechanics is the basis for many contemporary advanced constitutive modells describing the behaviour of soil.

Practising Engineers

Geotechnical engineers are typically graduates of a four-year civil engineering program and often hold a masters degree. In the USA, geotechnical engineers are typically licensed and regulated as Professional Engineers (PEs) in most states; currently only California and Oregon have licensed geotechnical engineering specialities.

State governments will typically license engineers who have graduated from an ABET accredited school, passed the Fundamentals of Engineering examination, completed several years of work experience under the supervision of a licensed Professional Engineer, and passed the Professional Engineering examination.

Soil mechanics

In geotechnical engineering, soils are considered a three-phase material composed of: rock or mineral particles, water and air. The voids of a soil, the spaces in between mineral particles, contain the water and air.

The engineering properties of soils are affected by four main factors: the predominant size of the mineral particles, the type of mineral particles, the grain size distribution, and the relative quantities of mineral, water and air present in the soil matrix. Fine particles (fines) are defined as particles less than 0.075 mm in diametre.

Soil Properties

Some of the important properties of soils that are used by geotechnical engineers to analyse site conditions and design earthworks, retaining structures, and foundations are:

Unit Weight: Total unit weight: Cumulative weight of the solid particles, water and air in the material per unit volume. Note that the air phase is often assumed to be weightless.

Porosity : Ratio of the volume of voids (containing air, water, or other fluids) in a soil to the total volume of the soil. A porosity of 0 implies that there are no voids in the soil.

Void ratio : is the ratio of the volume of voids to the volume of solid particles in a soil. Void ratio is mathematically related to the porosity.

Permeability : A measure of the ability of water to flow through the soil, expressed in units of velocity.

Compressibility : The rate of change of volume with effective stress. If the pores are filled with water, then the water must be squeezed out of the pores to allow volumetric compression of the soil; this process is called consolidation.

Shear strength : The shear stress that will cause shear failure.

Atterberg Limits: Liquid limit, plastic limit, and shrinkage limit. These indices are used for estimation of other engineering properties and for soil classification.

Geotechnical Investigation

Geotechnical engineers perform geotechnical investigations to obtain information on the physical properties of soil and rock underlying (and sometimes adjacent to) a site to design earthworks and foundations for proposed structures, and for repair of distress to earthworks and structures caused by subsurface conditions. A geotechnical investigation will include surface exploration and subsurface exploration of a site. Sometimes, geophysical methods are used to obtain data about sites. Subsurface exploration usually involves in-situ testing (two common examples of in-situ tests are the standard penetration test and cone penetration test. In addition site investigation will often include subsurface sampling and laboratory testing of the soil samples retrieved. The digging of test pits and trenching (particularly for locating faults and slide planes) may also be used to learn about soil conditions at depth. Large diametre borings are rarely used due to safety concerns and expense, but are sometimes used to allow a geologist or engineer to be lowered into the borehole for direct visual and manual examination of the soil and rock stratigraphy.

A variety of soil samplers exist to meet the needs of different engineering projects. The standard penetration test (SPT), which uses a thick-walled split spoon sampler, is the most common way to collect disturbed samples. Piston samplers, employing a thin-walled tube, are most commonly used for the collection of less disturbed samples. More advanced methods, such as ground freezing and the Sherbrooke block sampler, are superior, but even more expensive.

Atterberg limits tests, water content measurements, and grain size analysis, for example, may be performed on disturbed samples obtained from thick walled soil samplers. Properties such as shear strength, stiffness hydraulic conductivity, and coefficient of consolidation may be significantly altered by sample disturbance. To measure these properties in the laboratory, high quality sampling is required. Common tests to measure the strength and stiffness include the triaxial shear and unconfined compression test.

Surface exploration can include geologic mapping, geophysical methods, and photogrammetry; or it can be as simple as an engineer walking around to observe the physical conditions at the site. Geologic mapping and interpretation of geomorphology is typically completed in consultation with a geologist or engineering geologist.

Geophysical exploration is also sometimes used. Geophysical techniques used for subsurface exploration include measurement of seismic waves (pressure, shear, and Rayleigh waves), surface-wave methods and/or downhole methods, and electromagnetic surveys (magnetometre, resistivity, and ground-penetrating radar).

Foundations

A building's foundation transmits loads from buildings and other structures to the earth. Geotechnical engineers design foundations based on the load characteristics of the structure and the properties of the soils and/or bedrock at the site. In general, geotechnical engineers: 1) Estimate the magnitude and location of the loads to be supported; 2) Develop an investigation plan to explore the subsurface; 3) Determine necessary soil parametres through field and lab testing (e.g., consolidation test, triaxial shear test, vane shear test, standard penetration test); 4) Design the foundation in the safest and most economical manner.

The primary considerations for foundation support are bearing capacity, settlement, and ground movement beneath the foundations. Bearing capacity is the ability of the site soils to support the loads imposed by buildings or structures. Settlement occurs under all foundations in all soil conditions, though lightly loaded structures or rock sites may experience negligible settlements. For heavier structures or softer sites, both overall settlement relative to unbuilt areas or neighbouring buildings, and differential settlement under a single structure, can be concerns. Of particular concern is settlement which occurs over time, as immediate settlement can usually be compensated for during construction. Ground movement beneath a structure's foundations can occur due to shrinkage or swell of expansive soils due to climatic changes, frost expansion of soil, melting of permafrost, slope instability, or other causes. All these factors must be considered during design of foundations.

Many building codes specify basic foundation design parametres for simple conditions, frequently varying by jurisdiction, but such design techniques are normally limited to certain types of construction and certain types of sites, and are frequently very conservative.

In areas of shallow bedrock, most foundations may bear directly on bedrock; in other areas, the soil may provide sufficient strength for the support of structures. In areas of deeper bedrock with soft overlying soils, deep foundations are used to support structures directly on the bedrock; in areas where bedrock is not economically available, stiff "bearing layers" are used to support deep foundations instead.

Shallow Foundations

Shallow foundations are a type of foundation that transfers building load to the very near the surface, rather than to a subsurface layer. Shallow foundations typically have a depth to width ratio of less than 1.

Footings

Footings (often called "spread footings" because they spread the load) are structural elements which transfer structure loads to the ground by direct areal contact. Footings can be isolated footings for point or column loads, or strip footings for wall or other long (line) loads. Footings are normally constructed from reinforced concrete cast directly onto the soil, and are typically embedded into the ground to penetrate through the zone of frost movement and/or to obtain additional bearing capacity.

Slab Foundations

A variant on spread footings is to have the entire structure bear on a single slab of concrete underlying the entire area of the structure. Slabs must be thick enough to provide sufficient rigidity to spread the bearing loads somewhat uniformly, and to minimize differential settlement across the foundation. In some cases, flexure is allowed and the building is constructed to tolerate small movements of the foundation instead. For small structures, like single-family houses, the slab may be less than 300 mm thick; for larger structures, the foundation slab may be several metres thick.

Slab foundations can be either slab-on-grade foundations or embedded foundations, typically in buildings with basements. Slab-on-grade foundations must be designed to allow for potential ground movement due to changing soil conditions.

Deep Foundations

Deep foundations are used for structures or heavy loads when shallow foundations cannot provide adequate capacity, due to size and structural limitations. They may also be used to transfer building loads past weak or compressible soil layers. While shallow foundations rely solely on the bearing capacity of the soil beneath them, deep foundations can rely on end bearing resistance, frictional resistance along their length, or both in developing the required capacity. Geotechnical engineers use specialised tools, such as the cone penetration test, to estimate the amount of skin and end bearing resistance available in the subsurface.

There are many types of deep foundations including piles, drilled shafts, caissons, piers, and earth stabilised columns. Large buildings such as skyscrapers typically require deep foundations. For example, the Jin Mao Tower in China uses tubular steel piles about 1m (3.3 feet) driven to a depth of 83.5m (274 feet) to support its weight.

In buildings that are constructed and found to undergo settlement, underpinning piles can be used to stabilise the existing building.

Lateral Earth Support Structures

A retaining wall is a structure that holds back earth. Retaining walls stabilize soil and rock from downslope movement or erosion and provide support for vertical or near-vertical grade changes. Cofferdams and bulkheads, structures to hold back water, are sometimes also considered retaining walls. The primary geotechnical concern in design and installation of retaining walls is that the retained material is attempting to move forward and downslope due to gravity. This creates soil pressure behind the wall, which can be analysed based on the angle of internal friction and the cohesive strength (c) of the material and the amount of allowable movement of the wall. This pressure is smallest at the top and increases toward the bottom in a manner similar to hydraulic pressure, and tends to push the wall forward and overturn it. Ground water behind the wall that is not dissipated by a drainage system causes an additional horizontal hydraulic pressure on the wall.

Gravity Walls

Gravity walls depend on the size and weight of the wall mass to resist pressures from behind. Gravity walls will often have a slight setback, or batter, to improve wall stability. For short, landscaping walls, gravity walls made from dry-stacked (mortarless) stone or segmental concrete units (masonry units) are commonly used.

Earlier in the 20th century, taller retaining walls were often gravity walls made from large masses of concrete or stone. Today, taller retaining walls are increasingly built as composite gravity walls such as: geosynthetic or steel-reinforced backfill soil with precast facing; gabions (stacked steel wire baskets filled with rocks), crib walls (cells built up log cabin style from precast concrete or timber and filled with soil or free draining gravel) or soil-nailed walls (soil reinforced in place with steel and concrete rods).

For reinforced-soil gravity walls, the soil reinforcement is placed in horizontal layers throughout the height of the wall. Commonly, the soil reinforcement is geogrid, a high-strength polymer mesh, that provide tensile strength to hold soil together. The wall face is often of precast, segmental concrete units that can tolerate some differential movement. The reinforced soil's mass, along with the facing, becomes the gravity wall. The reinforced mass must be built large enough to retain the pressures from the soil behind it. Gravity walls usually must be a minimum of 30 to 40 percent as deep (thick) as the height of the wall, and may have to be larger if there is a slope or surcharge on the wall.

Cantilever Walls

Prior to the introduction of modern reinforced-soil gravity walls, cantilevered walls were the most common type of taller retaining wall. Cantilevered walls are made from a relatively thin stem of steel-reinforced, cast-in-place concrete or mortared masonry (often in the shape of an inverted T).

These walls cantilever loads (like a beam) to a large, structural footing; converting horizontal pressures from behind the wall to vertical pressures on the ground below. Sometimes cantilevered walls are buttressed on the front, or include a counterfort on the back, to improve their stability against high loads. Buttresses are short wing walls at right angles to the main trend of the wall. These walls require rigid concrete footings below seasonal frost depth. This type of wall uses much less material than a traditional gravity wall.

Cantilever walls resist lateral pressures by friction at the base of the wall and/or passive earth pressure, the tendency of the soil to resist lateral movement.

Basements are a form of cantilever walls, but the forces on the basement walls are greater than on conventional walls because the basement wall is not free to move.

Excavation Shoring

Shoring of temporary excavations frequently requires a wall design which does not extend laterally beyond the wall, so shoring extends below the planned base of the excavation.

Common methods of shoring are the use of sheet piles or soldier beams and lagging. Sheet piles are a form of driven piling using thin interlocking sheets of steel to obtain a continuous barrier in the ground, and are driven prior to excavation. Soldier beams are constructed of wide flange steel H sections spaced about 2–3 m apart, driven prior to excavation. As the excavation proceeds, horizontal timber or steel sheeting (lagging) is inserted behind the H pile flanges.

In some cases, the lateral support which can be provided by the shoring wall alone is insufficient to resist the planned lateral loads; in this case additional support is provided by walers or tie-backs. Walers are structural elements which connect across the excavation so that the loads from the soil on either side of the excavation are used to resist each other, or which transfer horizontal loads from the shoring wall to the base of the excavation. Tie-backs are steel tendons drilled into the face of the wall which extend beyond the soil which is applying pressure to the wall, to provide additional lateral resistance to the wall.

Earth Structures

Compaction: Compaction is the process by which the strength and stiffness of soil may be increased and permeability may be decreased. Fill placement work often has specifications requiring a specific degree of compaction, or alternatively, specific properties of the compacted soil. In-situ soils can be compacted either by excavation and recompaction, or by methods such as deep dynamic compaction, vibrocompaction, or compaction grouting.

Slope Stability

Slope stability is the analysis of soil covered slopes and its potential to undergo movement. Stability is determined by the balance of shear stress and shear strength. A previously stable slope may be initially affected by preparatory factors, making the slope conditionally unstable. Triggering factors of a slope failure can be climatic events can then make a slope actively unstable, leading to mass movements. Mass movements can be caused by increases in shear stress, such as loading, lateral pressure, and transient forces. Alternatively, shear strength may be decreased by weathering, changes in pore water pressure, and organic material.

Marine Geotechnical Engineering

In subsea geotechnical engineering, seabed materials are considered a two-phase material composed of 1) rock or mineral particles and 2) water. Structures may be fixed in place in the seabed—as in piers, jettys, or fixed-bottom wind turbines—or may be floating structures anchored to remain in a sea-surface position that remain roughly fixed relative to its geotechnical anchor point.

Undersea Foundations

Examples of undersea foundations include multiple-pile foundations as used in many piers and monopile foundations used for many fixed-bottom offshore wind turbines.

Floating-moored Structures

Undersea mooring of human-engineered floating structures include a large number of offshore oil and gas platforms and, since 2008, a few floating wind turbines.

Two common types of engineered design for anchoring floating structures include tension-leg and catenary loose mooring systems. "Tension leg mooring systems have vertical tethers under tension providing large restoring moments in pitch and roll.

Catenary mooring systems provide station keeping for an offshore structure yet provide little stiffness at low tensions." A third form of mooring system is the *ballasted catenary* configuration, created by adding multiple-tonne weights hanging from the midsection of each anchor cable in order to provide additional cable tension and therefore increase stiffness of the above-water floating structure.

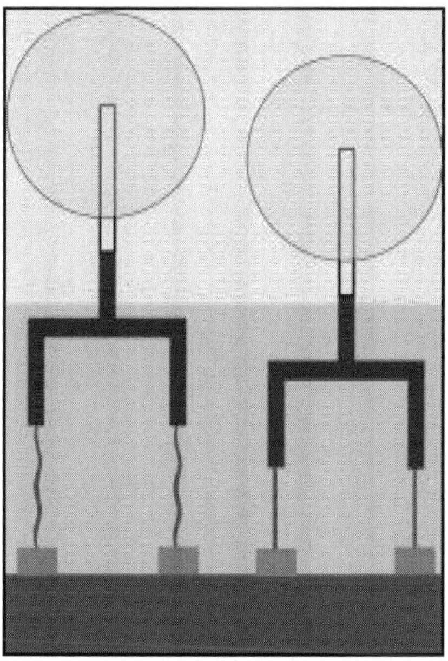

Figure : *A tension leg mooring system for a wind turbine: left-hand tower-bearing structure (grey) is free floating, the right-hand structure is pulled by the tensioned cables (red) down towards the seabed anchors (light-grey)*

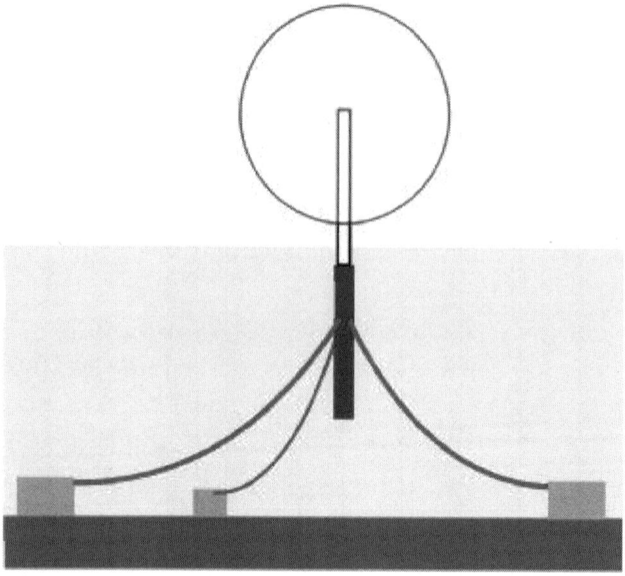

***Figure :** A floating wind turbine moored by catenary cables tied to fixed foundation points on the seafloor*

Geosynthetics

Geosynthetics is the umbrella term used to describe a range of synthethic products used to aid in solving some geotechnical problems. The term is generally regarded to encompass four main products; geotextiles, geogrids, geomembranes, and geocomposites. The synthetic nature of the products make them suitable for use in the ground where high levels of durability are required, though this is not to say that they are indestructible.

Geosynthetics are available in a wide range of forms and materials, each to suit a slightly different end use. These products have a wide range of applications and are currently used in many civil and geotechnical engineering applications including roads, airfields, railroads, embankments, retaining structures, reservoirs, canals, dams, landfills, bank protection and coastal engineering.

Bibliography

Anbuvelan, K. : *Management Concepts for Civil Engineers*, Laxmi Publications, Delhi, 2005.

Armstrong, J. A.: *Forces Shaping the U.S. Academic Engineering Research Enterprise*, National Academy Press, Washington DC, 1995.

Armytage, W. H. G.: *A Social History of Engineering*. Cambridge: MIT Press, 1966.

Basu, S.K. : *Global Dictionary of Civil Engineering*, Global Vision, Delhi, 2012.

Beaty H. Wayne: *Standard Handbook for Electrical Engineers*. New York, NY: McGraw-Hill, 2000.

Benjamin Henry: *The Engineering Drawings of Benjamin Henry Latrobe,* Yale University Press, NY, 1980.

Bennett, S.: *A History of Control Engineering: 1800-1930*, IEE Press, London, 1979.

Bhavikatti, S. S. : *A Textbook on Elements of Civil Engineering and Engineering Mechanics*, New Age International , Delhi, 2011.

Booker, P. J.: *A History of Engineering Drawing*, Chatto & Windus, London, 1963.

Calhoun, Daniel H. : *The American Civil Engineer: Origins and Conflict*. Cambridge: MIT Press, 1960.

Chen, W.F.: *Civil Engineering Handbook*. Boca Raton, FL: CRC Press, 1995.

David Fisher: *Rules of Thumb for Engineers and Scientists*, Gulf Pub, Houston, 1991.

Davison, C. S. C. B.: *Engineering Heritage*, Institute of Mechanical Engineers, London, 1963.

Domminghaus, H.: *Plastics for Engineers: Materials, Properties, Applications*, Hanser Publishers, Munich, 1993.

Dym, J.B.: *Product Design with Plastics: A Practical Manual*, Industrial Press, New York, 1983.

Edward H. Smith: *Mechanical Engineer's Reference Book,* Butterworth-Heinemann, Oxford, 1994.

Emerson, Howard P.: *Origins of Industrial Engineering: The Early Years of a Profession.* Institution of Industrial Engineers, 1988.

Fernandez, Martin : *Dictionary of Civil Engineering*, A P H, Delhi, 2011.

Fitchen, John: *Building Construction Before Mechanization.* MIT Press, 1986.

Florman, S.: *The Civilized Engineer*, St. Martin's Press, New York, 1979.

Groover, Mikell P.: *Fundamentals of Modern Manufacturing*, John Wiley & Sons, INC, New York, 2002.

Gupta, J.P. : *A T B of Engineering Mech. and Basic Civil Engineering*, S. Chand Publishing, Delhi, 2011.

Hasan, Syed Danish : *Civil Engineering Materials and Their Testing*, Narosa, Delhi, 2011.

Howard B., Cary,: *Modern Welding Technology*, Pearson Education, Chicago, New Jersey, 2005.

Joshi, Jagat Pati *: Harappan Architecture And Civil Engineering*, Rupa Pub, Delhi, 2008.

Kapoor, R : *Encyclopedia of Civil Engineering*, SBS Pub, Delhi, 2007.

MacCollum, David V.: *Construction Safety Planning.* New York, NY: Van Nostrand Reinhold, 1995.

Mendel, Otto: *Practical Piping Handbook.* Tulsa, OK: PennWell Books, 1981.

Nicholas P.: *Handbook of Chemical Engineering Calculations*, McGraw-Hill, New York, 2004.

Punmia, B.C.; Ashok Kr. Jain and Arun Kr. Jain: *Basic Civil Engineering*, Laxmi Publications, Delhi, 2003.

Richard Valentine: *Motor Control Electronics Handbook,* McGraw-Hill, New York, 1998.

Robert C. Rosaler: *Standard Handbook of Plant Engineering,* McGraw-Hill, New York, 1995.

Seely, Bruce E: *Building the American Highway System: Engineers as Policy Makers.* Philadelphia: Temple University Press, 1987.

Singh, Gurcharan : *Standard Handbook of Civil Engineering*, Standard Pub, Delhi, 2003.

Thomas Sixsmith & R. Hanselka: *Handbook of Thermoplastic Piping System Design.* Marcel Dekker Ltd, London, 1997.

Weman, Klas: *Welding Processes Handbook*, CRC Press, New York, 2003.

Index

❑❑❑